Microbial Biotechnology for Sustainable Agriculture, Horticulture & Forestry

Microbial Biotechnology for Sustainable Agriculture, Horticulture & Forestry

D. JOSEPH BAGYARAJ
Formerly, Professor & Head
(Department of Agricultural Microbiology)
ICAR Emeritus Scientist & INSA Senior Scientist

University of Agricultural Sciences, GKVK Campus
Bangalore, Karnataka, India

New India Publishing Agency
Pitam Pura, New Delhi-110 088

Published by
Sumit Pal Jain *for*

New India Publishing Agency

101, Vikas Surya Plaza, CU Block, L.S.C. Mkt.,
Pitam Pura, New Delhi-110 088, (India)
Ph.: 011-27341616, Fax: 011-27341717, Mob.: 09717133558
E-mail: info@nipabooks.com
Web: www.nipabbooks.com

ISBN : 978-93-80235-82-0

Typeset at : Harminder *for* Laxmi Art Creation
Printed at : Jai Bharat Printing Press, Delhi

सत्यमेव जयते

डॉ. एस. अय्यप्पन
सचिव एवं महानिदेशक
Dr. S. Ayyappan
Secretary & Director General

भारत सरकार
कृषि अनुसंधान और शिक्षा विभाग एवं
भारतीय कृषि अनुसंधान परिषद्
कृषि मंत्रालय, कृषि भवन, नई दिल्ली–110 114

Government of India
Department of Agricultural Research & Education
and Indian Council of Agricultural Research
Ministry of Agriculture, Krishi Bhawan, New Delhi-110 114
Tel.: 23382629, 23386711, Fax: 91-11-23384773
E-mail: dg-icar@nic.in

Foreword

Green revolution in India during 1970's no doubt brought about self-sufficiency in food production. However, excessive use of inorganic fertilizers and plant protection chemical for maximizing crop yields resulted in deterioration of physical, chemical and biological health of cultivated land. Hence, now we talk about sustainable agriculture. Sustainability refers to productive performance over time. It is the successful management of resources for agriculture to satisfy changing human needs while maintaining or enhancing the quality of the environment and conserving natural resources. Sustainability in agricultural production has emerged as one of the most significant concerns in the present century. Recent studies have shown that soil biodiversity plays an important role in determining agricultural sustainability. Use of chemicals is not only deleterious to environment, but also injurious to human, animal, plant and soil health. Under these circumstances, alternate sources of fertilizers and pesticides are looked for.

Organic wastes and microorganisms are alternate sources to meet the nutrient requirement and to control pests, diseases and weeds of crop plants to bridge the future needs. The quick generation time of microorganisms and the ease with which their nuclear material can be handled, make them ideal for biotechnologists/microbiologists to tailor them to yield desired products to serve mankind. Thus, microorganisms can be harnessed to produce more biofertilizers, to decompose organic wastes more efficiently and combat plant diseases and pests with greater efficiency than even before. Use of such microorganisms as inoculants or their products may help sustainable productivity not only in agriculture and horticulture but also in forestry.

Viewing from the above stated perspective, the book on 'Microbial Biotechnology for Sustainable Agriculture, Horticulture and Forestry' by Dr. D.J. Bagyaraj, NASI Senior Scientist Platinum Jubilee Fellow, is very timely. The book is comprehensive in coverage with 10 chapters. It covers all areas of microbially mediated processes including nutrient cycles in nature, mobilization of nutrients to plant growth, plan disease and pest control, decomposition of organic wastes, and technologies used for mass production of these organisms along with their quality control specifications. The book will be a valuable companion to students, teachers, researchers, administrators and policy-makers who foresee the need for non polluting, less-expensive and renewable sources to generate food, fibre and fuel for feeding mankind in the years to come.

Dated the 7, January, 2011 **(S. Ayyappan)**
New Delhi

Preface

Microorganisms are every where. Microorganisms can be found in the air we breathe, in the food we eat, and even within our own body. In fact, we come in contact with countless number of microorganisms everyday. Because microbes affect our everyday lives, microbiology provides many challenges and offers many rewards. Biotechnology is the product of interaction between the science of biology and technology. It is defined as "the controlled use of biological agents such as microorganisms or cellular components for beneficial use". Biotechnology plays a very important role in agriculture as a result of its ability to modify microorganisms, plants, animals, and agriculture processes. The umbrella of microbial biotechnology in agriculture covers numerous scientific activities, ranging from production of biofertilizers to that of microbial pesticides; from biological nitrogen fixation to ligno-cellulose degradation; from production of biomass and bio-fuels to transgenic plants.

Sustainability refers to productive performance of a system over time. It implies use of natural resource to meet the present needs without jeopardizing the future potential. A sustainable system must be sustainable both ecologically and economically. Most soils of the tropics are low in inherent fertility. Currently there is considerable resistance against the use of chemical pesticides and fertilizers; because of their hazardous influence on the environment, and on soil, plant, animal and human health. Hence the current day emphasis is on use of organic inputs including microbial inoculants which play an important role in sustainable agriculture, horticulture and forestry. Some books have been published on different groups of microorganisms used as biofertilizers, biocontrol agents and for composting of agricultural wastes, but there are a very few books bringing different aspects of microbial technology for enhancing plant growth and yield. Hence, the object of this book is to cover comprehensively different groups of microorganisms used for sustained productivity of plants important in agriculture, horticulture and forestry.

The book is organized into 10 chapters. Chapter 1 gives an introduction to the subject highlighting historical developments in the field. Chapter 2 helps to understand the physico-chemical characteristics of soil and covers the different groups of soil microorganisms, nutrient cycles and soil-plant-microbe interactions. Chapter 3 deals with various types of nitrogen fixing organisms including the physiology of nitrogen fixation. Chapter 4 is devoted to microorganisms modulating phosphorus nutrition of plants like phosphate solubilizing microorganisms and mycorrhizal fungi. Chapter 5 deals with plant growth promoting rhizomicroorganisms. Chapters 6 and 7 provide up-to-date information on microorganisms used for the biocontrol of soil-borne plant pathogens and insect-pests respectively. Chapter 8 discusses the role of microorganisms for control of weeds. Chapter 9 covers the role of microorganisms in the conversion of organic wastes to compost. Chapter 10 provides an overview of microbial biotechnology used in the mass production of microbial inoculants including their quality control specifications.

The readers may appreciate the potential and fascination of biotechnological approaches used for utilizing microorganisms in sustainable agriculture, horticulture and forestry. The book is a comprehensive and detailed analysis of the subject. However, I do not claim it to be an exhaustive study. Those who are eager to delve deeper in to the subject may consult the references at the end of each chapter. I am certain that the book will be useful to students, teachers and researchers interested in microbiology, biotechnology, natural resource management, organic farming and sustainable agriculture, horticulture and forestry.

D.J. Bagyaraj

Acknowledgement

This book was catalyzed and supported by a grant from the Department of Science and Technology (DST), Government of India under its Utilization of Scientific Expertise of Retired Scientists (USERS) Scheme, which I wish to gratefully acknowledge. My special thanks to Dr. B. Harigopal and Dr. S. S. Kohli of DST, New Delhi for their encouragement and support to undertake this work.

I also wish to thank all those who have assisted, in one way or the other, in bringing out this book. Special thanks are due to Messrs M. Srinivasan, R. Suresh, M. Raghavendrakumar, Mrs. B. Kavitha Nair and Drs. Hemlata Chauhan, B. Ramanujam, R. Rangeswaran, P. Sreerama Kumar, M.S. Rao, S.S. Hussaini. I also wish to record my deep sense of gratitude for M/s. New India Publishing Agency, New Delhi, for taking all pains and interest for timely and excellent publication. I will be failing in my duty if I do not say a word of thanks to my family members, especially my wife Mrs. Indrani Bagyaraj and my son Mr. J. E. Ravi, who helped me in many ways for the success of this venture.

D.J. Bagyaraj

Contents

Chapter-1

Introduction and Historical Perspectives

Biotechnology plays a very important role in agriculture as a result of its ability to modify microorganisms, plants, animals, and agricultural processes. The umbrella of microbial technology in agriculture covers numerous scientific activities, ranging from production of biofertilizers to that of microbial pesticides; from biological nitrogen fixation to lignocellulose degradation; from production of biomass and biofuels to transgenic plants. Terrestrial ecosystem contains an important dynamic living matrix termed soil, which is critical for agricultural production and food security. Soil forms the medium for growth and maintenance of a variety of plant species. Soil is considered a store house of microbial activity, though the space occupied by living microorganisms is estimated to be less than 5% of total space.

There is enough evidence in the literature to believe that microorganisms were the earliest of the living things that existed on this planet. It has been estimated that the origin of the universe can be traced back to about 5,000 million years. Owing to the physical, chemical and organic evolutions which took place since then, the earth with its various biological systems came into existence in its present form. Recent studies in paleontology have shown that the earliest form of life on earth was a unicellular alga, formed about 2,000 million years ago. From this bicellular, the multicellular and pleomorphic living being got evolved. The role of microorganisms in the evolutionary processes on earth is therefore very important. Their role as a source of food for other forms of the life and in various other physiological and biochemical processes on earth has become more significant in recent years. It is obviously, therefore, that microorganisms, which formed the first biological system on earth, had played and continue to play a pivotal role in the evolutionary process on this planet, thus, the extent to which man depends on microorganisms for his survival on earth cannot be overemphasized.

Man depends on crop plants for his existence and crop plants, in turn, depend on microorganisms for their nutrition. It is believed that even the pre-historic man cultivated crop plants some 10,000 years ago for making his food. There are evidences in the carvings of caves in Palestine to show that man, prior to 2,500 B.C., had tried to improve crops by ploughing the soil. Since then there have been attempts to improve the soil for plant growth by various means. Ancient Greeks and early Romans seem to have devoted great attention to improve their agricultural practices, from time to time, from the seventeenth century onwards, up to the first century B.C. The Romans seem to have attached particular significance to crop rotation with cereals and legumes, though the significance of microbes in atmospheric nitrogen fixation in such a rotation was not known until several centuries later.

Early Romans thought that legume plants enrich the soil, and this view was held for several centuries, though its scientific basis was not known. The accumulation of nitrates in soil as a result of the decomposition of organic matter was known in the seventeenth and eighteenth centuries. The importance of composting stable manure to obtain nitrate for plants were well understood by the French and English who undertook a campaign for the purpose soon after the French revolution. By the middle of the nineteenth century it was recognized that in order to make organic wastes more useful to plants, they should be composted, but the microbial role in decomposition was not understood until after its discovery by Pasteur.

In 1983, J.B. Boussingault opined that legumes fixed atmospheric nitrogen in soil though he did not find any proof. W.B. Leishman in 1858 and M.S. Woronin in 1866 demonstrated that the nodules in legumes were formed by a specific group of bacteria. Some considered them as pathogens infecting the roots. In 1856, J. von Liebig showed that nitrates were formed in soil due to addition of nitrogenous fertilizers, but he did not know the real cause of nitrification in soil. Pasteur was of the opinion that the nitrification in soil was a bacterial process. He also established that decomposition depended upon the nature of organic matter and the environmental conditions under which it was decomposed by the organisms. In 1877, T. Schloesing and A. Muntz, in their studies on purification of sewage water, found evidence for the role of bacteria in nitrification. R. Warington in 1878 demonstrated that nitrification in the soil is a microbial process, and this took place in two stages, the first stage being conversion of ammonia to nitrite and the second, conversion of nitrite to nitrate. In 1886 H. Hellriegel and H. Wilfarth in their classical experiment showed that the growth of non-leguminous plants was directly proportional to the amount of nitrogen supplied where as in legumes there was no relationship between the quantity of nitrogen supplied and the extent of plant growth. They suggested that bacteria in the root nodules of legumes accumulated atmospheric nitrogen and made it available for plant growth. However, it was S.N. Winogradsky, who in 1890, demonstrated beyond doubt the role of bacteria in nitrification. He isolated two groups of organisms and showed through experiments their nitrifying capacity, using carbondioxide as the carbon source for their cultivation. In 1892, T. Schloesing and M. Laurent showed that the weight of nitrogen absorbed from the air by the bacteria was approximately equal to the gain by the plant and the soil. Together, these various studies brought out the importance of nitrogen fixing bacteria in legume nodules and their contribution to scientific agriculture.

With the discovery of the gelatin plate technique by Koch in 1882, several studies were initiated to examine the specific types of bacteria in soil. In most of these studies, attempts were made by the scientists to coordinate the activities of the soil microorganisms with various chemical processes taking place in the soil and its fertility levels, and their efforts were a great success. It was established by S.N. Winogradsky in 1891 and by M.W. Beijerinck in 1893 and 1901 that the transformation of nitrogen in nature is largely due to the activities of various groups of soil organisms. Ammonification of organic nitrogenous substance by microorganisms was studied in detail by J.G. Lipman and P.E. Brown in 1903. Starting from the early years of the present century, emphasis was laid on the study of various types of microorganisms in soil, their physiology, interrelationship, and significance in soil processes and fertility levels. At times, the complexity of soil population and their interrelationships led to confusion in the minds of scientists. Protozoa were observed to consume bacteria, and hence they were thought to be harmful to soil fertility. Soon this was found erroneous as the soil microbial population was seen to remain in the dynamic equilibrium. The role of fungi and actinomycetes in the cellular decomposition in soil came to be better understood. Various other physiological groups like the sulphur oxidizing, anaerobic cellulose decomposing microorganisms, and iron bacteria were isolated from soil and their properties studied in detail. The publication of book *Principles of Soil Microbiology* in 1927 by Selman A. Waksman gave a fillip to the branch of microbiology, and since then, numerous investigators have conducted specialized studies in soil microbiology. Again in 1942, Waksman showed the importance of the soil as the source of antagonistic organisms, which led to extensive studies on soil actinomycetes and better understanding of the physiological and biochemical properties. Today soil is considered to be the main source scavenging the organic wastes through microbial action and is also a rich storehouse for industrial microflora of great economic value.

Though L. Hiltner in 1904 coined the term *rhizosphere* to denote the region around the root surface with increased microbial activity, interest in the rhizosphere microflora gained importance in 1960s. Several attempts were made to artificially influence the rhizosphere microorganisms for the benefit of plant growth during 1950 - 70s by R.L. Starkey, A.D. Rovira, G. Rangaswami and others.

Biological nitrogen fixation is another area of research which has attracted many scientists throughout the world. Techniques of isolation, screening and selection of efficient rhizobia for grain and forage legumes, factors affecting legume-*Rhizobium* symbiosis and methods of inoculation are currently being studied by the rhizobiologists. Rhizobial inoculants are now being produced industrially in many parts of the world. In 1976, J. Doberiner of Brazil highlighted *Azospirillum* as associated symbiont in grasses, and since then considerable work has been done on classification, ecology, physiology and genetics of *Azospirillum* and other non-symbiotic nitrogen fixing bacteria. The work on nitrogen fixation in nodulated plants other than legumes was taken up by Bond in the UK, and Quispel and Becking in Europe who made considerable headway on this aspect in 1960's.

Dommergues and associates from France and Senegal have discovered the occurrence of nodules on stem of *Sesbania rostrata* which fix nitrogen by virtue of which, the legume serves as an excellent green manure in low-land rice cultivation. Callahan, Torry and their associates (1978) found BNF activity by an actinomycete *Frankia* which produced nodules in non-legume roots. Using sunlight, which is abundant in the tropics, the cultivation of nitrogen-fixing cyanobacteria and *Azolla,* a water fern harbouring the cyanobacterium *Anabaena,* has been practiced in Thailand, Vietnam, China and India. It is this kind of technology which can be taken as an example in planning future strategies in agriculture to save on costly inputs which come from non-renewable source of energy.

In 1885, B. Frank coined the term *mycorrhiza* to denote the association of certain fungal symbionts with plant roots. The work done by Barbara Mosse in Rothemsted and J.W. Gerdemann in Illinois on endomycorrhizae colonizing the agricultural crop plants, and that of James Trappe from Oregon and Don Marx from Georgia on ectomycorrhizae colonizing the forest species in the last few decades, has aroused interest in many workers throughout the world. Biological control of plant pathogens using antagonistic microorganisms is an area of research interest today, as the microbial preparations do not pollute the environment like chemical pesticides. The contributions of R.J. Cook, G.C. Papavizas, K.F. Baker and N.S. Kerr to this branch of study are worth-mentioning. The contribution of C.N. Acharya in 1940s towards better utilization of agricultural wastes for the production of biogas and compost, especially in developing countries, needs special mention. In 1970s R.J. Swaby from Australia developed a preparation, which he called *biosuper,* by mixing rock phosphate, sulphur and thiobacilli. The sulphuric acid produced by thiobacilli in the mixture dissolved the rock phosphate and there by enhanced phosphorus nutrition of plants. The Soviet microbiologists introduced a product called *phosphobacterin* in 1962, containing cells of *Bacillus megaterium* var. *phosphaticum* capable of solubilizing rock phosphate for seed and soil application. Martin Alexander in 1961 started a school of soil microbiology at the Cornell University, USA, with particular reference to microbiological aspects of pesticide degradation. Selman Waksman discovered the antibiotic *streptomycin* inhibiting pathogens, from the soil actinomycete *Streptomyces* sp.

With the advancement in molecular biology after the establishment of gene structure by Watson and Crick (1953) followed a series of discoveries. Ochoa and Kornberg (1965) demonstrated the method of RNA and DNA synthesis using *Azotobacter* and *Escherichia,* respectively, as tools in their studies. In mid 1970's , new and powerful set of technique known as "recombinant DNA technology", helped the soil microbiology to advance tremendously. Recombinant biotechnology has strong applications in the alleviation of biotic and abiotic stresses in soil, especially in the present era of climate change and global warming. The discoveries of monoclonal antibodies in 1975 by Kohler and Milstein; archaea in 1977 by Carl Woese and George Fox; methods for sequencing DNA in 1977 by Sanger, Niklen and Coulson; culturing of hyperthermophiles in 1982 by Karl Stetter; polymerase chain reaction (PCR) in 1982 by Kary Mullis; earth's smallest and most abundant photosynthesizing organism, *Prochlorococcus* in 1988 by Sallie Chisholm *et al*; the first genome sequenced for a cellular organism, *Haemophilus influenzae* in 1995 by Craig venter *et al*; the ribosomal structure obtained by X-ray diffraction in 2001 by Marat Yusupov *et al.* and public availability of the genome sequences of 1,000 bacteria in 2008 by National Centre for Biotechnology Information (USA) have helped soil microbiologists to work in different areas, such as biodegradation, microbial ecology, plant microbe interactions, environmental pollution and monitoring of genetically engineered microorganisms into the environment.

References

Alexander, M. 1977. *Introduction to Soil Microbiology* (2nd edn.), John Wiley & Sons, New York, USA.

Atlas, R.M. and Bartha, R. 1993. *Microbial Ecology: Fundamentals and Applications* (3rd edn.) Benjamin-Cummings, Redwood City, USA.

Brock, T.D. 1998. *Milestones in Microbiology.* American Society for Microbiology Press, Washington DC, USA.

Bryson, V. (ed.) 1959. *Microbiology-Yesterday and Today.* Institute of Microbiology, Rutgers, University Press, New Brunswick, N.J., USA.

Clark, F.E. 1977. *Soil Microbiology-It's a Small World.* Soil Science Society of America Pub., Madison, Wisconsin, USA. 41:238-41.

De Kruif, P. 1953. *Microbe Hunters.* Harcourt Brace & Co., New York, USA.

Gray, T.R.G. and Williams, S.T. 1971. *Soil Microorganisms.* Oliver & Boyd , Edinburgh, U.K.

Lechevalier, H.A. and Solotorovsky, M. 1965. *Three Centuries of Microbiology.* McGraw Hill, New York, USA.

Madigan, M.T., Martinko, J.M., Dunlap, P.V. and Clark, D.P. 2009. *Brock Biology of Microorganisms* (12th edn.) Pearson Benjamin Cummings, New York, USA.

Metting Jr.F.B. (ed.) 1993. *Soil Microbial Ecology.* Marcel Dekker Inc., New York, USA.

Paul, E.A. and Clark, F.E. 1989. *Soil Microbiology and Biochemistry.* Academic Press, New York, USA.

Rangaswami, G. and Bagyaraj, D.J. 1998. *Agricultural Microbiology* (2nd edn.) Prentice-Hall of India Pvt. Ltd., New Delhi, India.

Slonczewski, J.L. and Foster, J.W. 2009. *Microbiology – An Evolving Science.* W.W. Norton & Company, Inc., New York, USA.

Subba Rao, N.S. 1999. *Soil Microbiology.* Oxford & IBH Publishing Co., New Delhi, India.

Tate, R.L. III. 2000. *Soil Microbiology.* John Wiley & Sons, New York, USA.

Waksman, S.A. 1961. *Soil Microbiology.* John Wiley & Sons, New York, USA.

Chapter-2

Soil as a Reservoir for Agriculturally Important Microorganisms

The term soil refers to the outer, loose material of the earth's surface, a layer distinctly different from the underlying parent rock. A number of features characterize this region of the earth's crust. It is the natural medium, which supports plant growth. It provides the physical support needed for the root system and also supplies nutrients needed for plant growth. It is a perennial source of organic matter. It supports a vast population of microorganisms.

The processes involved in the formation of soil are slow, gradual and continuous and are the some total of environmental effects on rocks collectively known as the weathering of rocks. Rocks are exposed to sun, wind, and rain. The fluctuations in temperature cause cracks on the surface of rocks in which the minerals released by the action of rain water accumulate. The winds bring in particles of organic matter which settle on the cracks on the surface of rocks and later support the growth of lichens on the rocks. This is the beginning of the process of weathering which continues as a chain reaction resulting in the formation of soil. In nature this process may take hundred's or thousand's of years. The parent material from which the soil is formed has direct bearing on the fertility of soil. For example, soils derived from limestone are more fertile than those derived from granite.

The soil is composed of five major components - mineral matter, water, air, organic matter and a living population. Of the inanimate portion, the amount of mineral and organic matter is relatively fixed at a single site. The proportion of air and water, however, fluctuates. Air and water together account for approximately 50% of the soil volume, the volume so occupied representing the pore space. The mineral fraction, contributing generally around 45%, originates from the disintegration and decomposition of rocks. Organic matter is around 3 to 6% of the soil volume. The living portion of the soil which includes small animals and microorganisms contribute less than 1% of the total volume, yet it is undoubtedly essential for crop production and soil fertility.

There are six soil zones in the world and each soil zone may consist of many soil types. Each soil is characterized by the presence of different horizons which can be seen when a vertical section is cut into the soil, referred to as profile. The three major layers which make up the soil profile are the A, B and C horizons. Each horizon may have sub-divisions and may not be present in all soils. Transitional stages can exist between individual horizons. The A horizon is rich in organic matter and its decomposition products are through microbial activities. The B horizon has little organic matter, few plant roots and sparse microflora. In it, iron and aluminium compounds often accumulate. The A and B horizons represent the true soil. The C horizon at the bottom consists of weathering parent material (Fig. 1).

Different horizons

O	Organic debris
A	Organic matter undergoing decomposition
	Transition from B to A
B	Silicate clay minerals plus organic matter
	Transition from C to B
C	The weathered parent material
	Parent material (Solid rock)

Fig. 1 : Different horizons of a soil profile

Physical properties of the soil type depend on the size of the soil particles. Particles above 2.0 mm are generally classified as gravel or stones and others as sand (between 0.05 and 2.0 mm), silt (0.002 to 0.05 mm) and clay (less than 0.002 mm). The texture of soil depends on percentage of sand, silt and clay in it and is classified as clay, sandy clay, sandy loam and loamy sand etc. Soil particles occupy slightly more than half the space in soil, the remaining space between the particles, called the pore space, is occupied by water and air.

Chemical properties of soil reveal that it is made up of three main components namely organic matter, the derivatives of parent rock and the clay fraction. These three main components provide nutrients for plant growth. The organic matter is a potential source of N, P and S for plant growth. Microbial decomposition of organic matter releases the nutrients in organic residues in an available form from unavailable forms. The inorganic minerals come from derivatives of parent rock and are made available to plants by mineral decomposition. The clay fraction of soil provides secondary minerals and amorphous materials such as iron, aluminium, silicon, oxygen etc. Major gases in soil are N_2, O_2 and CO_2. The CO_2 content of soil air is 10 to 33 times more than that of the atmospheric air by volume. This high CO_2 content is because of the respiration of soil microorganisms leading to lower O_2 content of the soil atmosphere. In flooded rice soil methanogenic microorganisms produce methane which absorbs long wave radiation leading to global warming.

The physical and chemical characteristics of soil determine the nature of the environment in which microorganisms are found. These environmental characteristics in turn affect the composition of microscopic population both qualitatively and quantitatively. The microorganisms are generally found in the upper 15 cm of soil, but this can be affected by the soil management and other environmental factors. For a soil biologist, the soil environment is unique as it is a storehouse of several flora and fauna, which interact with each other leading to decomposition of organic matter, nutrition of agriculture crops, weathering of rocks etc.

Soil Microorganisms

Bacteria

Bacteria are the most dominant group of microorganisms in soil and probably equal one half of the microbial biomass in soil. They are present in all types of soil but their population decreases as the depth of the soil increases. In general, horizon A of a soil profile consists of more microorganisms than B and C horizons. Under anaerobic conditions (in the absence of oxygen), bacteria dominate the scene and carry on the microbiological activities in the soil since fungi and actinomycetes do not grow well in the absence of oxygen.

Bacteria live in soil as cocci (spheres, 0.5 μm), bacilli (rods, 0.5 to 0.3 μm) or spirilli (spirals) (Fig. 2). The bacilli are common in soil whereas spirilli are very rare in natural environments. In 1925, Winogradsky, classified soil microorganisms in general and bacteria in particular into two broad categories - the autochthonous and the zymogenous organisms. The autochthonous or indigenous population is always uniform and constant in soil since their nutrition is derived from native soil organic matter (examples, *Arthrobacter* and *Nocardia*). On the other hand, zymogenous fermentative organisms require an external source of energy and their normal population in soil is low (examples, *Pseudomonas* and *Bacillus*). When specific substrates are added to soil the number of zymogenous bacteria increases and gradually declines when the added substrate is exhausted. To this category belong the cellulose decomposers, nitrogen utilizing bacteria and those splitting ammonium into nitrate.

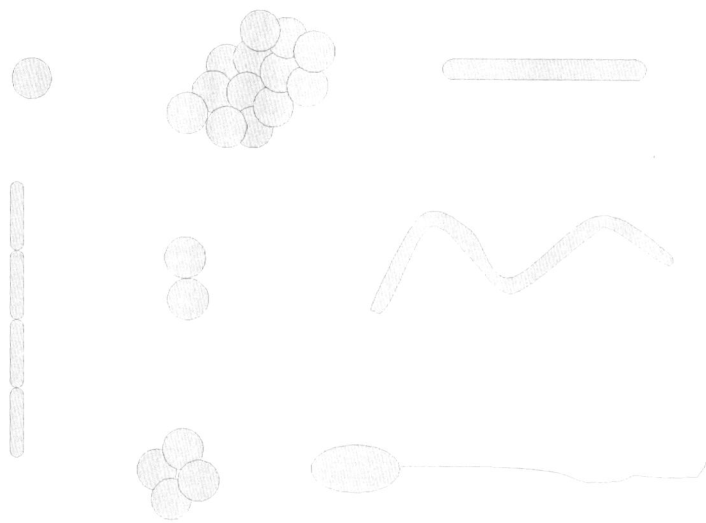

Fig. 2 : Some common bacteria occurring in soil

Autotrophic as well as heterotrophic bacteria are present in soil. Autotrophs synthesize their own food whereas heterotrophs depend on pre-formed food nutrition. Photoautotrophs are those whose food energy is derived through mediation of sunlight, as in the instance of photosynthetic bacteria as opposed to chemoautotrophs which oxidize inorganic materials to derive energy and at the same time utilize the carbon from CO_2 for growth. In the latter category, a group of bacteria known as obligate chemoautotrophs, are included which prefer specific substrates. Examples of this kind are *Nitrobacter*, which utilizes nitrite, *Nitrosomonas* which utilizes ammonium, *Thiobacillus* which converts inorganic sulphur compounds to sulphate and *Ferrobacillus* capable of converting ferrous iron to ferric iron. Several of the reactions involved in nitrogen transformations in soil depend on the chemoautotrophic *Nitrobacter* and *Nitrosomonas* and hence chemoautotrophy of bacteria in soil is intimately related to crop production

There are over 200 identified bacterial genera and a single soil sample may have over 4,000 genetically distinct bacteria. The most common soil bacteria come under the genera *Pseudomonas, Arthrobacter, Clostridium, Acromobacter, Micrococcus, Flavobacterium, Chromobacterium, Sarcina* and *Mycobacterium*. *Esherichia* is encountered rarely in soils except as a contaminant from sewage where as *Aerobacter* is frequently encountered and is probably a normal inhabitant of certain soils. Another group of bacteria common in soils is the myxobacteria belonging to the genera *Myxococcus, Chondrococcus, Archangium, Polyangium, Cytophaga* and *Sporocytophaga*. The latter two genera are cellulolytic and hence are dominant in cellulose rich environments. Myxobacteria feed on other Gram-negative bacteria through lysis.

It is not easy to determine the total population of bacteria in any soil accurately. Apart from the inherent limitations of the soil dilution and plate methods, their numbers vary with the texture, water content and many other parameters especially the availability of organic substrates in soil. Bacteria can withstand extremes of climate although temperature and moisture influence their population. They survive in a pH range from 2 to 8 and a temperature range from -5°C to 75°C. Counting bacteria in soil is difficult because no single culture media is adequate for all groups. About 10^8 to 10^{10} bacteria per gram of soil can be seen by direct observation. Growing bacteria on solid media usually gives 1% to 10% of the direct counts. Although numerous, bacteria represent less than 10% of the biomass in soil.

Actinomycetes

These are soil organisms, which have characteristics common to bacteria and fungi and yet possess sufficient distinctive features to delimit them into distinct category. In the strict taxonomic sense, actinomycetes are clubbed with bacteria. On agar plates, they can easily be distinguished from true bacteria. Unlike slimy distinct colonies of true bacteria which grow quickly, actinomycete colonies appear slowly, show powdery consistency and stick firmly to agar surface. A closer look at a colony under the compound microscope reveals slender unicellular branched mycelium (diameter of the hypha rarely exceeding one micron) forming asexual spores for propagation. They bear aerial mycelium which profusely sporulate and form distinct clumps or pellets in liquid cultures. Certain actinomycetes on the other hand, resemble *Mycobacterium* and *Corynebacterium* in all respects both morphologically and physiologically including susceptibility to the attack by viruses. Actinomycetes differ from fungi in the composition of their cell wall. They do not have chitin and cellulose which are commonly found in the cell walls of fungi.

The number of actinomycetes increases in the presence of decomposing organic matter. As a rule, they are intolerant to acidity and their numbers decline at pH 5.0. The most conducive range of pH is between 6.5 and 8.0. Waterlogging of soil is unfavorable for the growth of actinomycetes whereas desertic soils of arid and semi-arid zones sustain sizeable population, probably due to the resistance of spores to desiccation. The percentage of actinomycetes in the total microbial population increases with the depth of soil and actinomycetes can be isolated in sufficient numbers even from soil samples obtained from the C horizon of a soil profile.

In the order of abundance in soils, the commonest genera of actinomycetes are *Streptomyces* (nearly 70%), although *Nocardia, Micromonospora, Actinomyces, Actinoplanes* and *Streptosporangium* have also been encountered occasionally (Fig. 3).

Temperatures between 25 and 30°C are conducive for the growth of actinomycetes although thermophilic cultures growing between 55 and 65°C are common in compost heaps where they are numerically extensive and mostly belong to the genera *Thermoactinomyces* and *Streptomyces*.

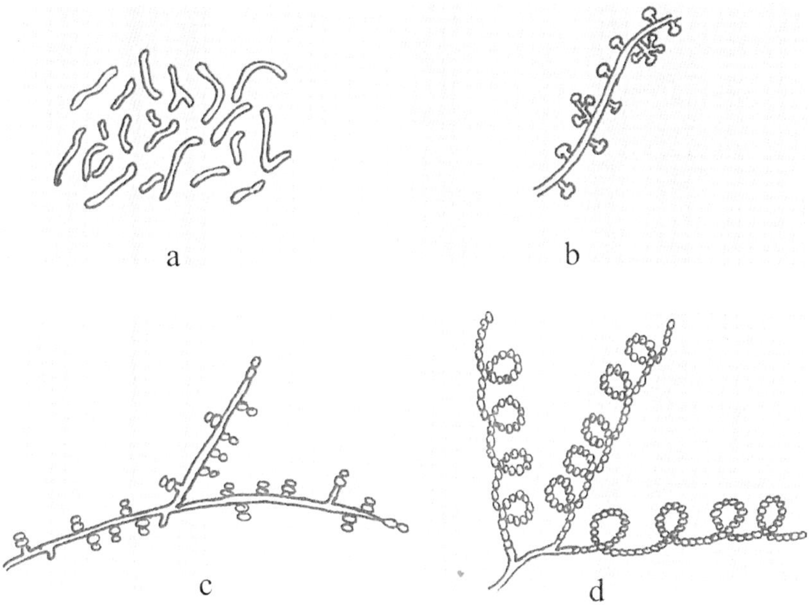

Fig. 3 : Some common actinomycetes occurring in soil
a. *Nocardia*, b. *Micromonospora*, c. *Microbispora*, d. *Streptomyces*

Fungi

Fungi are usually filamentous in contrast to most bacteria. In general, they are much larger than bacteria. In abundance they are next to bacteria in soil. The filamentous mycelia they possess are composed of individual hyphae.

The hyphae may be uni, bi, or multinucleate and non-septate (without cross walls) or septate. Asexual propagation by the production of spores or conidia takes place mitotically either with or without interception by a well defined sexual cycle involving gametic fusion and subsequent production of spores through meiotic or reduction division. The size, shape and colour of conidia or spores and the physiological characteristics of cultures in artificial as well as natural substrates provide valuable taxonomic criteria in the classification of fungal isolates into well defined genera and species.

All the environmental factors which influence the distribution of bacteria and actinomycetes also influence the fungal flora of the soil. The quality and quantity of organic matter present in soil have a direct bearing on fungal numbers in soil since most fungi are heterotropic in nutrition. Fungi are dominant in acid soils because acidic environment is not conducive for the existence of either bacteria or actinomycetes, resulting in the monopoly of fungi for utilization of native substrates in soil. They are also present in neutral or alkaline soils and some can tolerate pH beyond 9.0. Arable soils contain abundant fungi since they are strictly aerobic and excess of soil moisture decreases their numbers. Isolation of fungi from different horizons of soil profiles shows that these organisms exhibit selective preference for various depths of the soil. Those fungi which are common in lower depths are rarely encountered on the surface of soils which may be explained on the basis of the availability of organic matter and the ratio between oxygen and carbondioxide in the soil atmosphere at varying depths. Seasonal fluctuations in fungal numbers are not uncommon. Farm practices including crop rotation and fertilizer or pesticide application influence the nature and dominance of fungal species.

The following are the genera of fungi which are most commonly encountered in soils and which can be isolated by conventional methods: *Acrostalagmus, Aspergillus, Botrytis, Cephalosporium, Gliocladium, Monilia, Penicillium, Scopulariopsis, Spicaria, Trichoderma, Trichothecium, Verticillium, Alternaria, Cladosporium, Pullularia, Cylindrocarpon, Fusarium, Absidia, Cunninghamella, Mortierella, Mucor, Rhizopus, Zygorynchus, Pythium, Chaetomium,* and *Rhizoctonia* (Fig. 4).

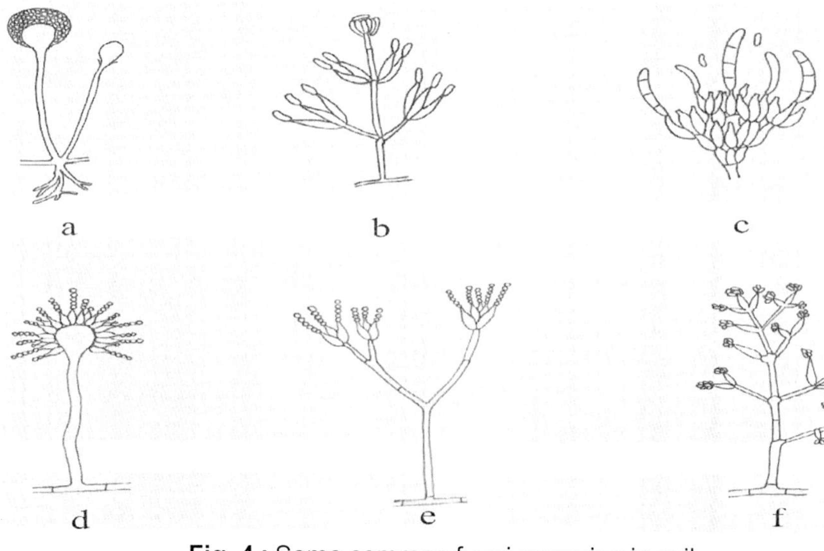

Fig. 4 : Some common fungi occurring in soil
a. *Rhizopus*, b. *Verticillium*, c. *Fusarium*, d. *Aspergillus*, e. *Penicillium*, f. *Trichoderma*

Many soil yeasts such as *Saccharomyces, Candida* etc. can also be isolated from soil on acidified media (pH 4.0). Their numbers in soil are low and their significance in soil is poorly understood.

One of the primary functions of filamentous fungi in soil is to degrade organic matter and help in soil aggregation. Besides this property, certain species of *Alternaria, Aspergillus, Cladosporium, Dematium, Gliocladium, Helminthosporium, Humicola* and *Metarhizium* produce substances similar to humic substances in soil and hence may be important in the maintenance of soil organic matter. Some of the fungi capable of forming ectotrophic associations on the root system of forest trees such as pine belonging to the genera *Boletus* and *Lactarius* help in the mobilization of soil phosphorus and nitrogen into plants. In many instances, establishment of new forest becomes difficult unless mycorrhizal fungi are artificially introduced in to soil by inoculation.

People also eat fungi. *Agaricus bisporus* (the button mushroom), *Pleurotus sojar-caju* (the oyster mushroom), *Calocybe* are few examples of edible fungi. These mushrooms are grown primarily on composed manure or paddy straw. Truffles and morels are two types of fungi highly priced by gourmets.

Algae

Algae are oxygen-evolving photosynthetic organisms that have "plantlike" chlorophylls and other photosynthetic pigments. They range in size from marine versions such as seaweed (kelp being an example) that form filaments 40 meters long, to microscopic species living in soil. *Chlorella*, for example is 2 to 3μm in diameter. Algae are the most common and wide spread plants. Although they are mainly aquatic, they inhabit all terrestrial environments. Algae are important in soils where light and water are available even though water may be only briefly present in these environments.

Algae have important ecological roles as pioneer colonizers in desert ecosystems where higher plants struggle to survive. Algae contribute to primary productivity, the formation of organic carbon compounds, and to the formation of soil structure. *Chlamydomonas,* a green alga, is added to sandy soils to aid erosion control and to clay soils because the extra cellular carbohydrates they produce improve aggregation and infiltration.

Algae associate with fungi to make complex organisms that are called lichens. Lichens contribute to silicate bioweathering by excreting organic acids. This is the first step in turning rocks into soil. Algae are near the bottom of the food chain. They are consumed by protozoa, nematodes, mites and earthworms. Consequently, pesticides and insecticides can stimulate algal blooms by eliminating predatory insect and microfaunal grazers.

There are 3 major groups of algae in soil, i) Green algae (Chlorophyta), ii) Diatoms, iii) Yellow-green algae (Xanthophyta) (Fig. 5).

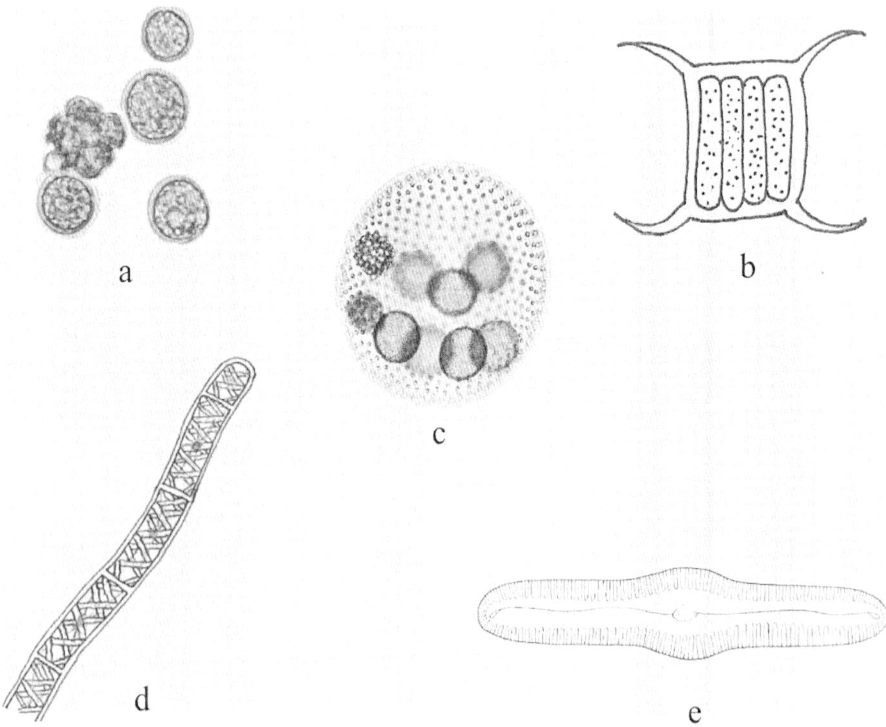

Fig. 5 : Some common algae occurring in soil
a. *Chlorococcum*, b. *Scenedesmus*, c. *Volvox*, d. *Spirogyra*, e. Diatom

Green algae contain chlorophyll *a* and *b* and have a cellulose cell wall that also contains other polysaccharides such as alginic acid. They may be unicellular or form colonies of many cells, or form long filaments of cells like beads on a string. Sexual reproduction is by fusion and asexual reproduction is by division (fission), zoospore formation, or fragmentation. Green algae store their photosynthetic products as starch, much like higher plants.

Diatoms contain photosynthetic pigments such as chlorophylls, carotenes, and xanthophylls. They store their photosynthetic products as oil. Diatoms are unicellular and reproduce sexually by fusion and asexually by fission and zoospore formation. They have a silica and pectin exoskeleton that resists degradation.

Yellow-green algae contain xanthophylls, carotenes, and chlorophyll. They have a pectin cell wall, although some also have a silica cell wall. Yellow-green algae are found as single cells and as cells that have formed filaments. They reproduce by fusion of cells and by fission of cells or by forming zoospores or simply fragmenting. The yellow-green algae store their photosynthetic products as oil.

Algae are less numerous than prokaryotes. Algal colony – forming units vary between 10^3 and 10^6 colony-forming units per gram of soil (Metting, 1993). Algae typically contribute between 7 and 300 kg of biomass per hectare. Algal populations vary, depending on soil type. In temperate soils, green algae are more numerous than diatoms, which in turn are more numerous than yellow-green algae. Green algae dominate in acid soils, while diatoms grow best in neutral soils. Algae are found in the upper layers of soil where there is both water and light. Temperature, light and water are the major factors controlling their population. Soil fertility has a direct effect on the abundance of algae. Since algae can produce their own carbohydrates, it is the lack of soil nutrients such as N, P, and K that help to control algal growth.

Protozoa

Soil protozoa are unicellular. In general they lack chlorophyll barring few exceptions. They are characterized by a cyst stage in their life cycle, which can help the species to withstand adverse soil conditions. Barring a few genera, which reproduce sexually by fusion of cells, the rest of them reproduce asexually by fission. The flagellated protozoans belonging to the class Mastigophora are predominant in soil. Important genera are *Allantion, Bodo, Cercobodo, Cercomonas, Entosiphon, Heteromita, Monas, Oikomonas, Sainouran, Spiromonas, Spongomonas, Euglena* and *Tetramitus*. Unlike flagellates, which move with the help of their flagella numbering up to four, there are protozoa in soil, which move with the help of extrusions of protoplasm known as false feet or pseudopodia. The protoplasm may be naked or as in some genera encased in shells. Important soil inhabitants of this class known as Sarcodina are *Amoeba, Biomyxa, Difflugia, Euglypha, Hartmanella, Lecythium, Naegleria, Nuclearia* and *Trinema*. A third group of soil protozoa belong to the class Ciliata. They are distinguished by the possession of minute hairs called cilia around their bodies, which help in locomotion. Important genera of this class are *Balantiophorus, Colpidium, Colpoda, Enchelys, Gastrostyla, Halteria, Oxytricha, Pleurotricha, Uroleptus, Paramecium* and *Vorticella* (Fig. 6). Apart from the classes Mastigophora, Sarcodina and Ciliata which are soil inhabitants, the phylum protozoa has two more classes (Suctoria and Sporozoa) which are not considered here.

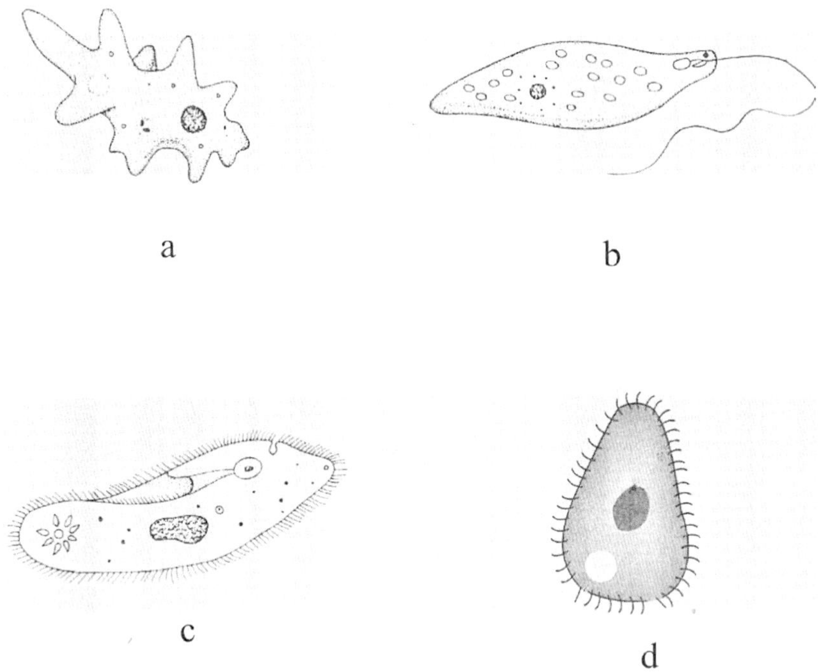

Fig. 6 : Some common protozoa occurring in soil
a. *Amoeba*, b. *Euglena*, c. *Paramecium*, d. *Colpoda*

Total protozoa in a soil may range from 10,000 to 100,000. There may be daily population changes in the protozoa. Fewer than 1,000 of this protozoa are ciliates; the rest are flagellates and sarcodina. Flagellates are the most numerous of the protozoan in a typical soil and they dominate acid soils. Protozoal biomass can be 5 to 20 g / m^2. Eliminating protozoa decreases the overall decomposition rate of organic matter in soil. This suggests that protozoa, like microfauna, help stimulate decomposition. Decomposition is faster in the presence of protozoa because microbial populations are kept young and active, even if individual members of the population get eaten.

Viruses

Viruses are "infectious agents made up of nucleic acids and proteins but unable to grow independently or reproduce by binary fission". Viruses are the smallest microorganisms, much smaller than bacteria. Most viruses are only 20 to 30 nm or smaller in size. They are obligate intracellular parasites. However, viruses can survive outside their host and remain infectious. Viruses are acellular organisms, lacking cell membrane. They contain RNA or DNA (the virion) surrounded by a protein coat (the capsid). Viruses come in many shapes. Some are helical, such as many plant viruses; some are cubical, based on a icohetradral pattern (a 20 sided shape), such as herpes viruses and some are binal, which means they have a cubical head with a helical tail, such as bacteriophages (viruses attacking bacteria) (Fig. 7). There are plant, animal, bacterial, fungal, algal and protozoal viruses. Viruses are extremely host specific. Little is known about the ecology of the viruses in soil. Plant viruses rarely survive in soil, while some insect viruses remain infective for years. Many soil-borne viruses require transmission by nematodes or fungi. There are reports that say productivity of alfalfa fields in France reduced because of bacteriophages infecting rhizobia forming root nodules and fixing atmospheric nitrogen (Angle, 1994). Viruses have been reported to attack more than 60 species of fungi. The most extensively studied system is the mycovirus of *Penicillium chrysogenum*. Accurate

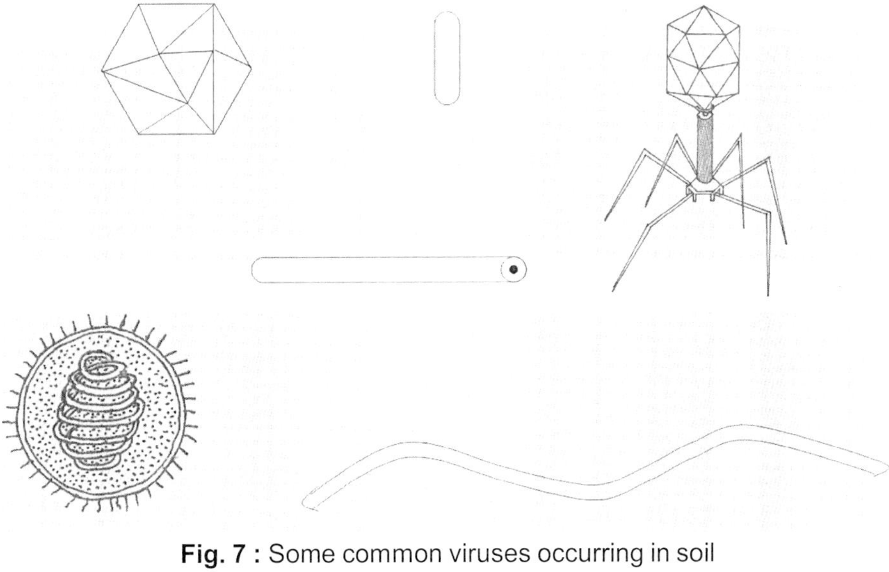

Fig. 7 : Some common viruses occurring in soil

determinations of molecular weight and amino acid content of different viruses have been made and electron microscope studies suggest that young apical regions of hyphae of *Penicillium* are free of virus particles whereas the older regions of hyphae contain many particles. Viruses have also been observed in sections of fungal spores. The relationship between fungal viruses and the metabolism or genetics of their hosts is poorly understood.

Factors Influencing Soil Microbial Population

The soil temperature, moisture and redox potential are the 3 major factors influencing microbial population in soil.

Soil temperature

The surface soil temperature in general ranges from 0 to 60°C. It is affected by angle of solar irradiation, soil shading, soil colour, soil depth and soil moisture. Most microorganisms in soil can grow over a temperature range of approximately 20-35°C (mesophiles). Some grow at greater than 90°C (thermophiles), and some grow at -10°C (cryophiles / psychrophiles). Most of the soil microorganisms are mesophiles growing over moderate temperatures.

Soil moisture

Soil moisture can greatly influence soil temperature. If the moisture is more the temperature will be less and *vice versa*. Most microorganisms prefer moisture between 20 to 60 %. Optimum microbial activity is at about -0.01 MPa and declines rapidly as the soil dries. In response to soil-drying, microorganisms have numerous strategies that include taking in external solutes and synthesizing internal solutes on demand or constitutively. The most resistant microorganisms to water stress are fungi, which can withstand water potentials as low as -40 MPa. Under dry soil conditions, the bacteria may form spores which can resist the drought conditions. The fungi may sporulate or form chlamydospores to tide over adverse conditions. Similarly, the protozoa may form cysts which can survive under dry conditions. It is believed that the availability of nutrients in soil is affected to a larger extent by the moisture content. If there is less of moisture, the concentration of nutrients in soil water is more, whereas when there is more moisture the concentration of nutrients is lesser. Also, at very low moisture levels the soil water may be bound to clay particles, and so they may not be useful to the microorganisms. On the other hand, when there is excess water, not only the nutrients are diluted, but also the aeration is greatly limited, and hence only the anaerobic and microaerophilic organisms can develop better.

Soil pH

Regarding pH, neutral pH *i.e* pH 7 supports maximum microbial population in soil. Acidic pH generally supports fungi, while neutral to alkaline pH support bacteria and actinomycetes. Exceptions always exist, and in the most extremely acidic conditions one may find bacteria rather than fungi. For example, *Thiobacillus desulphuricans* can grow in environments with a pH around 0.6. This environment exists around ore-leaching facilities or hot sulphur springs in volcanic environments.

Soil organic matter

Soil organic matter varies considerably in different soil types, from less than 0.5 per cent in desert soils to 40 per cent in peaty soils. The organic matter being the chief source of energy and food for most soil organisms, it has great influence on the population. The nature of organic matter is responsible for the differential stimulation of the population. The formation of humus, and its influence on soil and its microbes are discussed later. There are several indirect effects of the organic matter on soil microflora. It influences the structure and texture of soil, besides enriching it with nutrients for plants and microorganisms. Such influences on the soil also greatly affect the activity of the soil microorganisms.

Reduction-oxidation (Redox) potential

A redox reaction is the transfer of electrons from one compound (A) to another compound (B). During this reaction, A is oxidized (loses electrons) and B is reduced (gains eletrons). Microorganisms gain energy by oxidizing reduced material. The elements commonly involved in redox reactions are H, C, N, O, S, Fe and Mn. Redox potential (E_h) describes the tendency of electrons to flow between compounds. Electrons flow from electron-rich, low - E_h, reduced compounds to electron-poor, high- E_h, oxidized compounds. Redox potential reflects an electrical potential-the potential to drive electrons (current; so E_h is measured in volts) from reduced to oxidized compounds. The more positive the E_h, the greater the oxidizing capacity. Microorganisms conserve or harness that electron flow to generate energy.

Redox potential in soil is governed by the strongest oxidizing agent present. Oxygen is a very strong oxidizing agent; it has a tremendous affinity for electrons. As long as the soil is well-aerated, O_2 is at a high concentration and keeps the E_h high, which means that compounds will tend to be oxidized-to give their electrons to O_2. The ultimate electron acceptor is usually O_2. However, microorganisms also use NO_3^-, Fe^{3+}, Mn^{4+}, and SO_4^{2-} as electron acceptors. Any environment that has many microorganisms that are actively respiring and consuming O_2 can rapidly become anaerobic. This leads to anaerobic microsites in otherwise aerobic soils.

Agricultural practices

There are atleast two kinds of effects due to crop plants on the microorganisms. The plants excrete chemical substances through their roots which stimulate microbial growth. When the roots and other plant parts are left over in the field, they form rich nutrients for the microorganisms. The type of chemicals excreted by the plants and the selective stimulation of various soil organisms differ considerably. The organic waste added to the soil also varies, and depending upon the plant species added and its chemical composition, different organisms are stimulated in the soil. Crop rotation with different species like legumes, graminaceous plants, etc. brings about different stimulatory effects on the microflora. Some crops have deeper roots than others and some are more fibrous. Such variations bring about physical changes in soil which in turn may have direct and indirect stimulatory effects on the soil microflora. Various cultural practices, such as tillage operations, and irrigation, have several physical and chemical effects on soil, which are reflected on the soil microflora. Weeding the field may remove certain types of plant roots, and allow only the specific action of the roots of the crop plants on the microflora.

Irrigating the soil has several indirect effects on soil microflora. If the soil is waterlogged as in rice culture, the anaerobic and microaerophilic forms will be encouraged. If the soil dries, forming cracks in summer months, the population in the deeper soil layers will increase due to increased aeration. Application of fertilizers to the soil improves the microbial activity because of the availability of more readily obtainable nutrients. Some fertilizers, however, have inhibitory effect on specific bacterial types. Addition of nitrites inhibits the activity of free-living nitrogen-fixing bacteria like *Azotobacter*. Application of green manures or green leaf manures has many effects on soil microflora. First, these manures stimulate some bacteria, as also actinomyces and fungi. Spraying crop plants with pesticides, plant growth regulators, etc. has some effect on the soil microflora. The chemicals may fall on the soil and act directly to alter the population. On the other hand, the sprays may influence the plant growth and the root exudates may have indirect influence on the soil microflora. Thus every change, direct or indirect, in crop production may result in altering the soil microflora. But all these changes seem to be only temporary as very often the microbial equilibrium is restored to the original level.

Nutrient Cycles

The nutrient cycles such as carbon, nitrogen, phosphorus, sulphur, iron etc. routinely take place in the ecosystem. The productivity of plants and activity of animals including humans are greatly influenced by these cycles. Without soil microorganisms there could hardly be any nutrient cycling. The different nutrient cycles are briefly discussed below.

Carbon cycle

The maintenance of the organic constituents of the soil depends upon the photosynthetic capacity of the crop plants grown in the soil. The carbon cycle occurring in nature is illustrated in fig. 8.The photosynthesis of carbon compounds by green plants and photosynthetic bacteria helps in maintaining this balance. The organic substances synthesized by them serve as a source of energy and food material for other bigger as well as smaller forms of life.

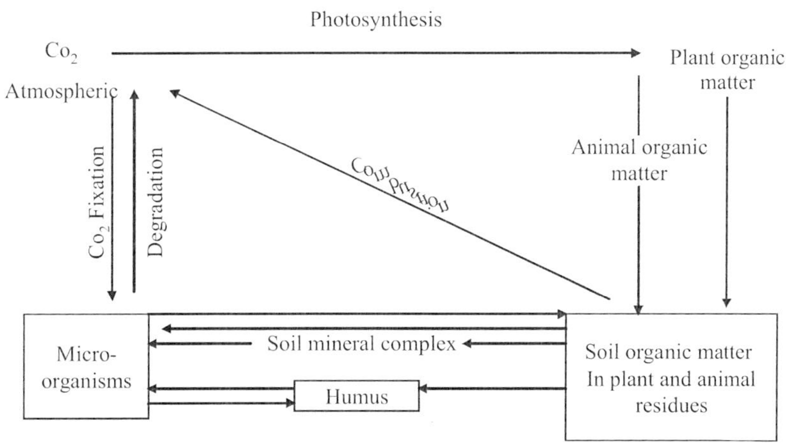

Fig. 8 : An illustration of carbon cycle

Some of these biological units, *viz.* animals and microorganisms, serve as food for other biological systems. A majority of these activities take place on earth particularly in the soil. The soil microorganisms play an important role in completing the cycle. They convert organic matter into body substances, liberate CO_2 and water, increase and concentrate the nitrogen content, and bring down the ratio between carbon and nitrogen in soil. This improves soil fertility. The released CO_2 goes back to complete the cycle and this ensures that there is no major lock-up in organic tissues on earth.

Humus

The plant and animal residues are attacked by soil microorganisms and are decomposed. In this process, some substances are volatilized, some used as food material by the microorganisms, others go into soil water, yet others are gradually transformed into humus. Humus is a complex organic residue left in the soil. The nature of this residue depends upon the composition of organic matter added, the microflora acting on them, and the soil conditions in which the humus is formed. To the mass of resistant residues of plant and animal matter in soil, the newly synthesized microbial cell substances and lysed cell walls are also added. Because of the constant addition and removal, the humus in soil is in a dynamic equilibrium. Humus is highly colloidal and occurs in soil as a colloidal complex with clay particles. The complex carries a negative charge and exhibits base exchange properties. On account of this property, the humus-clay complex is helpful in releasing nutrient elements to the plants. It acts as an acid or base towards soil minerals and mobilizes them for plant use.

The chemical analysis of humus yields (i) a small percentage of water-soluble organic substances such as sugars and amino acids, (ii) a fraction of phenolic glycosides or tannins, (iii) a fraction of pentosans and organic phosphates, and (iv) a polyuronoid substance containing complex sugars. Humus is largely insoluble in water, but completely soluble in alkali, and partially in acids. The carbon content is about 60 per cent and nitrogen 3 to 6 per cent, and possesses a high base-exchange capacity and combines with various inorganic substances in soil. The C:N ratio in humus is approximately 10:1. Humus improves the physical condition of soil as it can imbibe water and swell, and thus improve the water holding capacity of the soil. It is the storehouse of some essential plant food materials as it combines in itself iron, calcium, phosphorus, magnesium, manganese, etc. The resistant organic complex is slowly acted upon by certain soil microorganisms, and the released substances are made available for plant growth. It forms certain compounds with soil minerals and increases the buffering effect. For all these reasons, humus is important in determining the fertility level of soil. It serves as a source for plant food materials when most other sources of supplies are exhausted. Humus tends to make soils more granular with better aggregation of soil particles, and opens up adhering soil particles, prevents leaching of water-soluble substances, and several such direct and indirect effects. The biological activity of the soil is also greatly influenced by the humus content. Humus favours better development of plant-root system, which in turn directly and indirectly improves soil microbial activity.

The C:N ratio in fresh plant material varies from 80:1 to 12:1, *i.e.* the plant material is rich in bound-energy and poor in nitrogen. The energy is utilized by microorganisms and CO_2 is released. With the rapid multiplication of microorganisms, the demand for nitrogen increases. The available inorganic nitrogen in the soil is utilized by the microorganisms, and the plants are deprived of their nitrogen supply. As humification proceeds, the C: N ratio narrows down, and in the final stages, with the accumulation of nitrogen and release of CO_2, the ratio may reach 10:1. Hence, whatever may be the C: N ratio of the organic matter added to the soil, the final C: N ratio tends to reach somewhere near 10:1. This ratio remains more or less stable in most soils.

Regarding the soil organisms responsible for the attack, the heterotrophic groups are most active in the early stages. Some soil fauna like termites also reduce cellulose. In acid soils with pH around 4, many fungi, prominent among them *Trichoderma, Aspergillus,* and *Pencillium,* are active and they attack cellulose. The basidiomycetous fungi such as *Marasmius, Ganoderma, Psalliotta* attack lignin. In the less acid and neutral conditions, bacteria become active. They attack cellulose and hemicellulose. Actinomycetes and some bacteria also attack lignin in the neutral range. Under anaerobic or partial anaerobic conditions, bacteria become more active than fungi or actinomycetes. Some specific groups of bacteria are particularly efficient in reducing carbon compounds to form lactic, butyric and acetic acids; methane, hydrogen and ammonia are also produced. Some bacteria utilize hydrogen and reduce CO_2 to produce methane.

If glucose is attacked by microorganisms under aerobic conditions, then

$$C_6H_{12}O_6 + 1^1/_2\ O_2 \longrightarrow C_6H_8O_7 + 2H_2O$$

citric acid

$$C_6H_{12}O_6 + 4^1/_2\ O_2 \longrightarrow 3C_2H_2O_4 + 3H_2O$$

$$C_6H_{12}O_6 + 6O_2 \longrightarrow 6CO_2 + 6H_20$$

If it is attacked anaerobically by yeasts and bacteria, then

$$C_6H_{12}O_6 \longrightarrow 2C_3H_6O_3$$

lactic acid

$$C_6H_{12}O_6 \longrightarrow 2C_2H_5OH + 2CO_2$$

$$C_6H_{12}O_6 \longrightarrow C_4H_8O_2 + 2CO_2 + 2H_2$$

butyric acid

If it is attacked anaerobically by fungi, then

$$C_6H_{12}O_6 + 2H \longrightarrow C_4H_6O_4 + C_2H_5OH + H_2O$$

Under aerobic conditions

$$C_2H_5OH + 2O \longrightarrow CH_3COOH + CH_2O$$

$$2CH_3COOH + O \longrightarrow C_4H_6O_4 + H_2O$$

An increase in the evolution of the CO_2, as a result of the decomposition of plant and animal residues added to the soil, or the soil humus, leads to an increase in CO_2 content of the soil atmosphere. This results in a rise in the hydrogen-ion concentration of the soil.

$$CO_2 + H_2O \longrightarrow H_2CO_3$$

$$H_2CO_3 \longrightarrow H^+ + HCO_3^-$$

The hydrogen-ion will interact with various soil minerals, especially phosphates and silicates, and make them more soluble and ensure their availability for plant growth:

$$Ca_3(PO)_4 + 2H_2CO_3 \longrightarrow 2CaHPO_4 + Ca\ (HCO_3)_2$$

$$Na_2SiO_3 + 2H_2CO_3 \longrightarrow 2NaHCO_3 + H_2SiO_3$$

$$H_2SiO_3 \longrightarrow H_2O + SiO_2$$

The CO_2, therefore, hastens the weathering process of rock constituents. CO_2 in soil also supplements CO_2 of atmosphere for plant growth. The microorganisms thus regulate the CO_2 in the atmosphere and in plants. If they do not act on organic matter, the limited supply of CO_2 in the atmosphere will be exhausted and green plants would cease to manufacture carbohydrates. If the microbes are more active, then all the organic matter will be reduced to CO_2, and the soil will not be quite fit for plant growth. Thus, the carbon cycle is largely maintained by the balanced action of microorganisms in soil.

Cellulose is the most abundant organic matter in nature. Several workers have studied in depth the decomposition of cellulose reaching the soil, and several organisms participating in the process have been isolated. Cellulose is a polysaccharide composed of glucose molecules linked together in a linear chain of -1, 4- linkage, and the molecular weight ranges from 200,000 to 2 million. Diverse groups of fungi such as *Aspergillus, Fusarium, Trichoderma, Chaetomium* and *Curvularia* actively degrade cellulose found in plant tissues reaching the soil. Among the bacteria, *Bacillus, Acromobacter, Cytophaga, Pseudomonas, Nocardia, Micromonospora* and *Cellulomonas* are important. Also, some species of *Streptomyces, Nocardia, Micromonospora* and *Streptosporangium* possess cellulolytic properties. Some anaerobic bacteria and thermophilic fungi have been found to be active cellulose degraders under certain environmental conditions. The enzyme involved in all these organisms is cellulase and is adaptive.

Cellulase splits the long chain cellulose to yield glucose or cellobiose, *i.e.* two molecules of glucose linked together at the end of the chain. Cellobiose in turn can be split by cellobiase to yield two molecules of glucose.

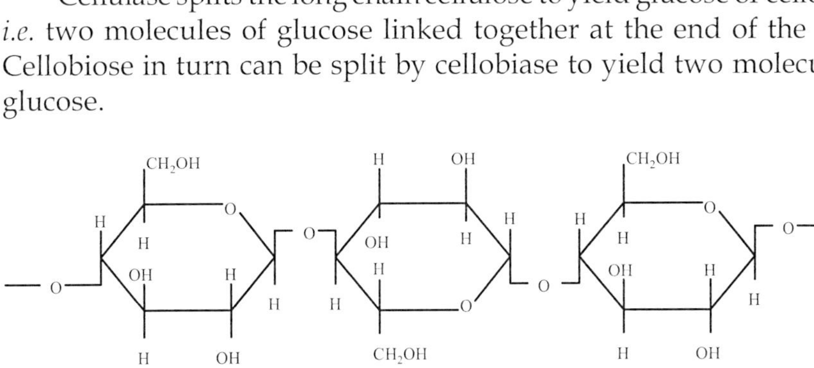

ß-1-4-glucoside structure of cellulose on degradation by cellulase enzyme gives rise to glucose molecules.

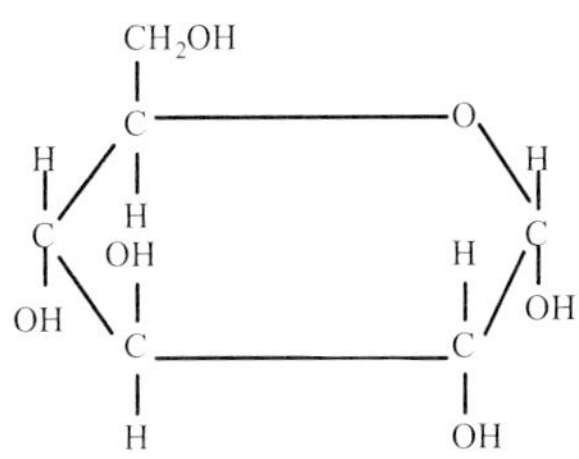

Glucose

Hemicelluloses are water-soluble polysaccarides and consist of hexoses, pentoses and uronic acids. Glucose, galactose, mannose, xylose, arabinose, glucuronic acid and galacturonic acid are commonly found in the hemicelluloses of plants. The hemicelluloses are more readily broken down by microorganisms than the other two. The molecule contains aromatic ring as the building block, with C_6 - C_3 type of linkage.

Microbial decomposition of lignin has attracted many investigations. Lignin is one of the most resistant organic substances for the microorganisms to degrade. Many basidiomycetous fungi have been found to possess special capacity in this regard. Only rarely have bacteria been found to reduce lignin. The fungi commonly found are *Fomes, Ganoderma, Agaricus, Armillaria, Pleurotus, Polyporus, Ustulina* etc. High lignin containing material, such as coir pith and coniferous wood, are preferentially attacked by the white-rot causing polyporous fungi. The enzymatic breakdown of lignins may result in simpler aromatic compounds such as syringaldehyde, vanillin, p-hydroxy-benzaldehyde, and ferulic acid. The final cleavage of the aromatic ring may take place with the involvement of other fungi and bacteria to yield low molecular weight organic acids, carbon dioxide, methane, etc.

Starch serves as a storage product of plants, and is present in several specialized parts such as tubers, bulbs, rhizomes, and seeds. It contains two components, *viz.*, amylose and amylopectin. In amylose the glucose molecules are linked together by an α-1, 4-glycosidic bondage, whereas in amylopectin, besides the same linkage of glucose molecules, side chains are attached through α-1, 6-glycosidic linkages. In starch molecule, amylopectin constitutes about 80 per cent and amylose about 20 per cent. Starch is more easily broken down than cellulose by soil microorganisms. Aerobically, the microbes fully utilize starch to produce carbon dioxide and low molecular weight organic acids. Anaerobically, fermantation takes place to yield methane, acetic, lactic and butyric acids. Bacteria, actinomycetes and fungi have the physiological capacity to utilize starch as a carbon source for growth and multiplication. When the soil is screened for starch hydrolyzers, the common organisms found include *Aspergillus, Fomes, Fusarium, Rhizopus, Streptomyces, Nocardia, Micromonospora, Bacillus, Chromobacterium, Flavobacterium, Micrococcus, Pseudomonas, Clostridium* and *Cytophaga*. The enzymes involved in the hydrolysis are α- and β-amylases and they act on 1, 4- linkage to yield maltose. Enzyme α-glucosidase in turn transforms maltose into glucose.

$$\text{Starch} \xrightarrow{\text{amylase}} \text{maltose} \xrightarrow{\alpha\text{-glucosidase}} \text{glucose}$$

Pectic substances are polysaccarides found in the constituents of middle lamella and in the primary and secondary cell walls of plants. They are made up of galacturonic acid units bound in a long chain. There are three types of pectic substances, *viz.* pectin, protopectin and pectic acid. Though closely related, each requires a specific enzyme to get hydrolysed.

Microorganisms which are capable of utilizing pectic substances as carbon and energy sources are abundant in soil and plant surface. Soil bacteria, such as *Bacillus, Clostridium* and *Pseudomonas,* and facultative pathogens like *Erwinia* more readily produce the pectinases. Several fungi, especially the plant pathogenic ones, and *Streptomyces,* are also known to produce the pectinases:

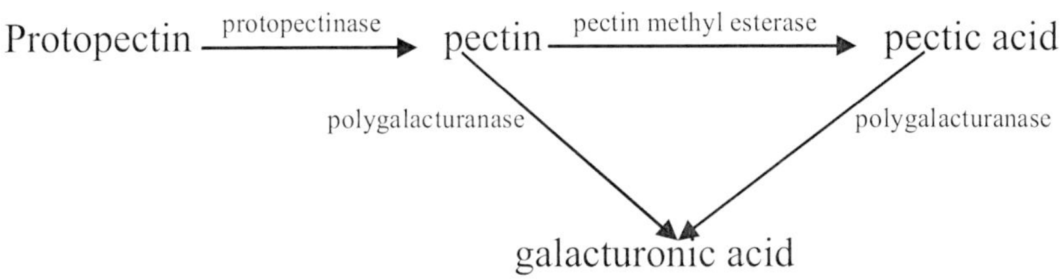

The enzymes involved are exoenzymes and the ability of the microorganism to produce them may vary. In the pathogenic organism, certain specificity for host range is conferred on its inherent capacity to produce one or more of the pectic enzymes.

Chitin is a polysaccharide whose basic unit is an amino sugar. It consists of a long chain with an empirical formula of $(C_6H_9O_4.NHCOCH_3)_n$. It is a structural component giving mechanical strength to several plants and animals. It is also present in fungal cell wall and in insects. This is one of the hardiest organic molecules for microbial action in soil. *Streptomyces,* and to some extent *Nocardia* and *Micromonospora,* readily decompose chitin. Among the fungi, some species of common soil fungi such as *Aspergillus, Penicillium, Mucor, Trichoderma* and *Thamnidium* also possess the capability to break down chitin. The breakdown products are glucosamine, acetic acid, ammonia and CO_2. Chitinase initiates the process to reduce chitin to diacetylchitobiose, and chitobiase reduces the latter into acetylglucosamine.

There are several other complex carbon compounds such as gum, inulin, and related substances present in plant and animal tissues, which eventually reach the soil. These substances are selectively attacked by soil fungi and bacteria, through their exoenzymes which are also adaptive enzymes. The long chains of the organic substances are broken down to smaller units, and subsequently through the action of other organisms, complete mineralization of the compounds takes place.

The protein content of the waste products, on hydrolysis by specific enzymes produced by many varieties of microorganisms, is split into polypeptides and finally into simpler amino acids. These amino acids are either utilized by the microorganisms for body build-up, or acted upon by some bacteria and fungi, producing ammonia, carbon dioxide, various organic acids, and alcohol. Under anaerobic conditions, various amino acids and mercaptans are formed, producing an offensive odour. Besides protein, other nitrogenous substances, which are added to the soil, are first converted into urea, which is again decomposed to ammonia and carbon dioxide. Oils and fats are hydrolysed to glycerol and fatty acids. Glycerol is readily oxidized to carbon dioxide and water, and fatty acids produce certain resistant and toxic substances. The transformation of fatty acids under anaerobic conditions is a process which is believed to have contributed to the formation of petroleum in nature. *Clostridium perfringens* has been shown to form combustible material from olive oil.

The rates at which these chemical changes take place in the soil vary according to the chemical nature of the organic matter added. Some of the organic constituents are attacked very easily, others less quickly, and still others are highly resistant to the attack. The plant materials like straw, stubbles and forest litter, which are very low in their nitrogen content, are more resistant to the attack by the microbes. When an organic waste is added to the soil, the more readily available constituents are decomposed quickly and the resistant residue left as a black mass. This mass, together with the dead microbial cells, forms a complex water-insoluble, but alkali-soluble substance called humus. Under favourable conditions, humus undergoes constant but slow decomposition by certain microbes, resulting in simpler substances. Humus thus forms the storehouse of essential elements required by the plants.

Nitrogen cycle

Some of the important groups of reaction falling under "nitrogen cycle" are now briefly described, and an outline is given in fig. 9. Proteolysis or decomposition of proteins by microorganisms takes place in many stages. Proteins consist of amino acid units linked together by the peptide bondage. They are invariably of L-optical form, and their molecular weight may be as high as 1,000,000. The structure of amino acid molecules varies from very simple forms to highly complicated molecules. They may combine to form small units, known as peptides. Many peptide units join to form protein molecules. These molecules are large and vary in different organic tissues. The protein content of such organic plant residues like rice straw may be less then 1 per cent, but legumes contain about 20 per cent protein. Animal residues and organic manures contain larger percentages of protein. The proteins are broken down by the microorganisms with the help of proteolytic enzymes. The breakdown of proteins is accomplished in two stages:

$$\text{Proteins} \xrightarrow{\text{proteinases}} \text{polypeptides} \xrightarrow{\text{peptidases}} \text{amino acids}$$

In the first stage, proteinases breakdown the large molecules of proteins into smaller units of polypeptides, each containing two or more amino acids. In the second stage, the polypeptides are broken down further into tripeptides and dipeptides, and then to free amino acids by the enzyme peptidases. The production of proteinases by microorganisms depends on several factors. It seems possible that, when protein is added to the soil, the fungi first attack the molecule and reduce it to polypeptides, and then bacteria become active and compete with fungi in further breaking down the polypeptides to amino acids. The amino acids may be utilized by other microorganisms for their growth, and may be passed to plants though mycorrhiza or may be further broken down to smaller molecules. This breakdown is accomplished by a number of ways by the microorganisms, but in most cases the product is ammonia. In this respect, bacteria are relatively more active than other organisms. Spore forming bacteria, pseudomonads, actinomycetes and fungi seem to readily attack amino acids in soil. However, in soils fungi are more important agents of ammonification than bacteria.

Amino acids are broken down by oxidative deamination by aerobic microorganisms.

$$\underset{\text{glutamic acid}}{COOH.CH_2CH_2CHNH_2.COOH} + O_2 \longrightarrow \underset{\text{succinic acid}}{COOH.CH_2.CH_2.COOH} + NH_3 + CO_2$$

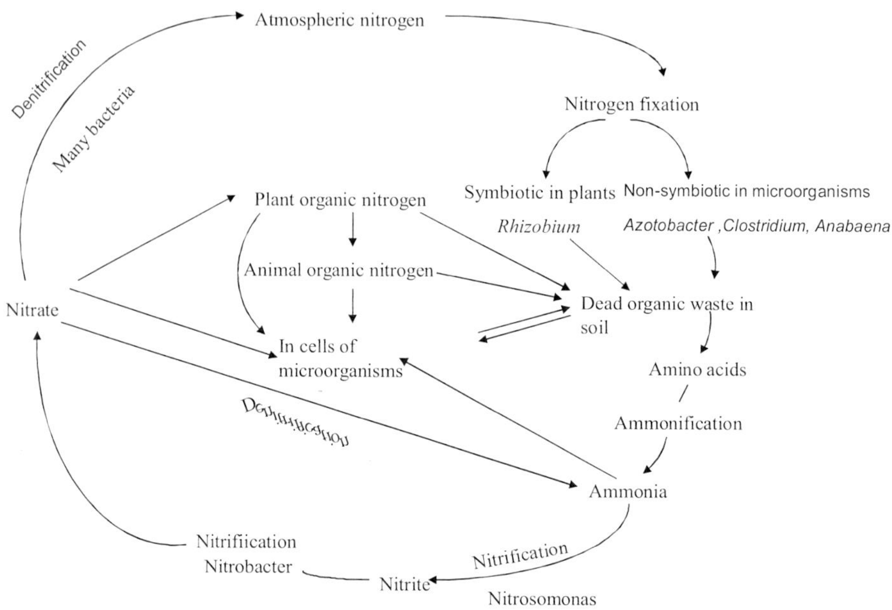

Fig. 9 : An illustration of nitrogen cycle

Thus from glutamic acid, succinic acid, ammonia and carbon dioxide are produced. Succinic acid may be made available for plant growth and CO_2 released to the soil atmosphere, and finally to the atmosphere above the soil. Bacteria break down complex nitrogenous substances to obtain energy, and in this process NH_3 is released.

Amino acids may be broken down by hydrolytic deamination:

$$R.CHNH_2.COOH + H_2O \longrightarrow RCHOH.COOH + NH_3$$

$$R.CHNH_2.COOH + H_2O \longrightarrow RCH_2OH + CO_2 + NH_3$$

$$R.CHNH_2.COOH + H_2 \longrightarrow R.CHO + HCOOH + NH_3$$

Thus, depending on the enzyme system involved in the hydrolytic process, we may obtain organic acids, alcohols, aldehydes, carbon dioxide, etc. besides ammonia. Sometimes, the amino acid may be decarboxylated in two steps:

$$CH_3.CHNH_2.COOH + H_2O \longrightarrow CH_3.CH_2NH_2 + CO_2$$

$$CH_3.CHNH_2.COOH + H_2O \longrightarrow CH_3CH_3 + NH_3 + CO_2$$

Under anaerobic conditions, reductive deamination may take place. The hydrogen requirement for this process may be obtained from sources in the soil complex.

The anaerobic bacteria may also produce ammonia as indicated below by the destructive deamination:

$$CH_3.CH_2.CHNH_2.COOH \longrightarrow CH_3CH{=}CH.COOH + NH_3$$

Several fungi are capable of producing alcohols and ammonia from amino acids:

$$\begin{matrix} CH_3 \\ CH_3 \end{matrix} \!> CH.CH.CHNH_2COOH \longrightarrow \begin{matrix} CH_3 \\ CH_3 \end{matrix} \!> CH.CH_2.CH_2.OH + NH_3 + CO_2$$

leucine — isoamyl alcohol

Thus, the amino acids are broken down to give rise to ammonia, CO_2 and a variety of substances such as lower organic acids, aldehydes, and ketones.

There are several factors influencing the ammonification of proteins in soil. In well aerated and drained soils there is more ammonification. If a large amount of organic matter, rich in protein is added, there is more of ammonification, whereas if organic residues such as cereal straw are added to the soil, there is rapid multiplication of organisms. These organic residues bind up, at least temporarily, all the available nitrogen and, therefore, nitrogen in the form of ammonia is not liberated. In acid soils, fungi are more active than bacteria, but the general microbial activity is relatively less, and hence there is less of ammonia production. If there is a ready supply of energy-rich carbohydrates, the microorganisms do not attack the complex nitrogen compounds. Thus, for ammonification, rich carbohydrates are harmful.

Besides proteins, other nitrogen containing organic substances are also attacked by microorganisms to yield ammonia. Many organisms utilize urea to liberate ammonia:

$$NH_2CONH_2 + 2H_2O \longrightarrow \underset{\text{ammonium carbonate}}{(NH_4)_2\,CO_3} \longrightarrow 2NH_3 + CO_2 + H_2O$$

Members of the genera *Bacillus, Proteus, Micrococcus, Sarcina, Aerobacter,* etc. are known to quickly convert urea added to soil to ammonium carbonate and then ammonia. The enzyme responsible for the utilization of urea is urease. Few bacteria are known to utilize amines as C and N sources for their growth. Species of *Mycobacterium, Pseudomonas, Protoaminobacter* etc. are known to possess capacities to utilize methylamine, ethylamine, propylamine, etc. These are the by-products of microbial breakdown of proteins. The enzyme systems involved in this process are amine oxidases:

$$NH_2.RCH_2NH_2 \longrightarrow NH_2.RCH = NH$$

$$NH_2.RCH = NH \longrightarrow NH_3 + NH_2.R.CHO$$

$$NH_2.R.COOH \longrightarrow CO_2 + NH_3$$

$$NH_2.R.CH_2OH \longrightarrow CH_3.CHO + NH_3$$

Thus, the products will be CO_2, NH_3, and H_2O. *Chorella* can utilize amides as nitrogen source.

Plants and animal residues contain nucleic acids in small proportions. When they reach soil, they are also attacked by soil microorganisms and are broken down as given in fig 10.

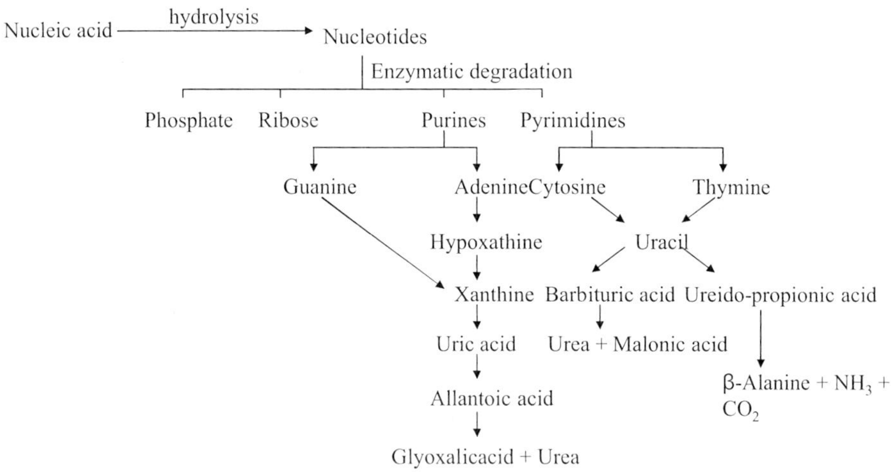

Fig. 10 : Breakdown of nucleic acid by microorganisms

The hydrolysis of nucleic acids by aerobic bacteria and some fungi is accomplished through nuclease. The breakdown of purines, pyrimidines has not been well understood. *Enterobacter aerogenes* and species of *Clostridium* are reported to attack some purines and pyrimidines. *Clostridium* spp. seem to attack purines and produce methane. Direct deamination of adenine by means of adenase has been reported in the case of *Escherichia coli*. Thus, by the breakdown of the various nitrogenous substances by the microorganisms, NH_3, CO_2, and other by products are obtained.

The ammonia produced by the reactions mentioned above may be utilized in several ways:

(i) It may be utilized by some organisms for their growth, *i.e.* mycorrhizal fungi are capable of utilizing ammoniacal nitrogen and make it available for plant growth.

(ii) It may be utilized by higher plants for their growth.

(iii) Ammonia may be oxidized to nitrite and then to nitrate (nitrification).

(iv) Under highly specialized conditions, there may be some volatalization of ammonia, resulting in a loss. This takes place when there is excess of nitrogen in the soil. When there is high temperature prevailing above the soil and when fungi vigorously attack urea, ammonia is liberated

However, some protein complexes which are resistant to microbial attack persist for a long period of time. Such proteins combine with humus, lignin, soil clay-particles, etc. The net result is that there is slow mineralization of nitrogen from the bound complex form. Such a mineralization of organic nitrogen is interlinked with the mineralization of organic carbon.

Nitrification

Though it has been known for long that decomposition of organic matter in soil result in increased concentration of nitrate in soil, it was only in 1877 that T.Schloesing and A.Muntz showed that the mechanism of conversion of ammonia to nitrate was biological by nature. This was further confirmed by R.Warrington in 1886, who showed that special bacteria are involved in the process. Calcium carbonate promoted nitrification and sometimes the conversion stopped half-way at nitrite. S. Winogradsky (1881-91) brought out the important characteristic of bacteria, namely their capacity to oxidize ammonia to obtain energy, which is similar to photosynthesis in plants. But instead of light as a source of energy, chemical oxidation may provide energy for the build-up processes of the microbial cell. Based on this principle, he isolated bacterium which oxidized ammonia to nitrite, and called it *Nitrosomonas,* and another bacterium which could convert nitrite to nitrate and named it *Nitrobacter.* Later, he isolated several other species of bacteria capable of oxidizing ammonia to nitrate. Winogradsky's studies gave impetus to the work of several investigators, and various autotrophic bacteria were isolated. Later, nitrifying bacteria which are active in acid soils in the presence of organic matter were also isolated by other workers.

Nitrobacter is inhibited in cultures by the presence of nitrite and *Nitrosomonas* by ammonia. Both are active only under aerobic conditions, and calcium carbonate improves their activity. The nitrification process taking place in soil through microorganisms is indicated by the following chemical reactions:

$$\underset{\text{ozone}}{(NH_3)_2 + 4O_3} \longrightarrow \underset{\text{ammonium nitrite}}{NH_4NO_2} + H_2O_2 + 4O_2$$

$$\underset{\text{ammonium nitrate}}{NH_4NO_2 + H_2O_2} \longrightarrow NH_4.NO_3 + H_2O$$

The chemical reaction of the biological oxidation may be represented as follows:

$$2NH_3 + 3O_2 \longrightarrow 2\ HNO_2 + 2H_2O$$

$$2NH_3 + 3O_2 \longrightarrow N_2O_3 + 3H_2O$$

$$NaNO_2 + \tfrac{1}{2}\ O_2 \longrightarrow NaNO_3$$

$$NH_4^- + \tfrac{1}{2}\ O_2 \longrightarrow NO_2^- + 2H^+ + H_2O$$

$$NH_2OH + O_2 \longrightarrow HNO_2 + H_2O$$

$$\underset{\text{ammonia}}{NH_3 + \tfrac{1}{2}\ O_2} \longrightarrow \underset{\text{hydroxylamine}}{HO\text{-}NH_2} \xrightarrow{-2H} \underset{\text{hyponitrite}}{\tfrac{1}{2}\ HO\text{-}N{=}N\text{-}OH} \xrightarrow{\tfrac{1}{2}O_2} \underset{\text{nitrite}}{HO\text{-}N{=}O}$$

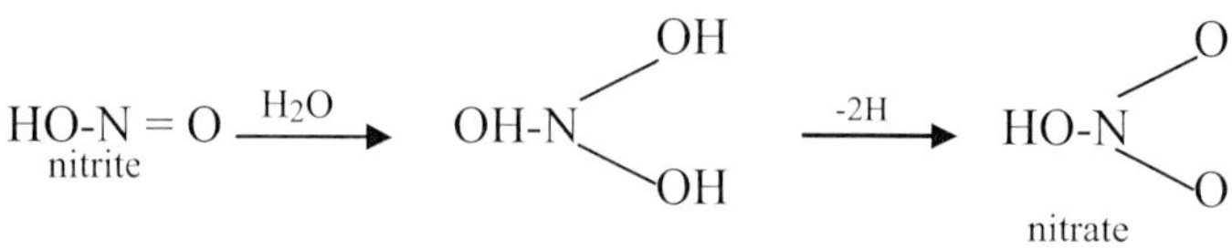

Though originally it was thought that organic matter can be nitrified directly, A. Muntz showed that ammonia should be first liberated from the organic matter, which is then oxidized to nitrite. W.L. Omeliansky further confirmed that all organic nitrogenous substances have to be converted to ammonia and then oxidized to nitrite and nitrate. Conditions such as well-aerated soil rich in calcium carbonate, a temperature below 30°C, nearly neutral pH, and the absence of large quantities of organic matter seem to favour nitrification in soil. Application of nitrate fertilizers reduces their activity, but phosphates and other minerals enhance their action. Ammonia is believed to be converted to nitrate in the presence of a catalyst at high temperature. It may be due to phytochemical action at the soil surface in the tropical countries, *i.e.* due to the effect of sunlight in the presence of certain photosynthetic substances like TiO_2 and ZnO, which act as catalysts. It would convert ammonia into nitrate, as claimed by N.R. Dhar and his students in India. The nitrate thus formed may be (i) utilized by other microorganisms, (ii) reduced to nitrite and ammonia or (iii) reduced to ammonia and nitrogen gas.

Plants assimilate large quantities of nitrate from soil for their growth. Acidity alters the uptake of nitrates by plants. The cation exchange is proportional to nitrification. The liberation of nitrate from humus is higher in acid soils than in saline and alkaline soils.

Denitrification

This is the reverse process of nitrification, *i.e.* nitrate is reduced to nitrites and then to nitrogen gas and ammonia. Unlike nitrification, denitrification process is favoured in anaerobic soils, and there is considerable decomposition or putrefaction of organic matter taking place. In some cases the denitrification process may be a part of assimilatory process of the organism. In the absence of aeration, *i.e.* lack of free oxygen supply, some organisms denitrify completely, while in other cases the process may not be complete. Anaerobic bacteria like *Clostridium* may reduce nitrate to ammonia directly:

$$HNO_3 + 4H_2 \longrightarrow NH_3 + 3\,H_2O$$

$$HCOOH + HNO_3 \longrightarrow CO_2 + HNO_2 + H_2O$$

$$2HNO_3 \longrightarrow 2HNO_2 + O_2$$

$$2HNO_2 \longrightarrow NH_2 + 1\tfrac{1}{2}\,O_2 + H_2O$$

$$HNO_3 + H_2 \longrightarrow NH_3 + 2O_2$$

$$5C_6H_{12}O_6 + 24HNO_3 \longrightarrow 24H_2CO_3 + 6CO_2 + 18H_2O + 12N_2$$

Some aerobic organisms such as *Pseudomonas denitrificans* also seem to reduce nitrate under certain conditions. The presence of denitrifying bacteria in soil is well established, and the presence of large quantities of organic matter is known to encourage the liberation of nitrogen into the atmosphere. The loss of nitrogen is more than in the case when high doses of organic matter are added together with inorganic nitrate fertilizers. When soil is waterlogged, the loss is more, and this is more so in hot climate. In waterlogged paddy fields, however, if nitrates are applied, nitrites which might prove toxic to the plants may be formed. Soon the nitrite may be attacked by other organisms and converted into ammonia and nitrogen. The enzyme required is nitrate reductase or nitratase. This enzyme has been detected in *Escherichia coli, Pseudomonas aeruginosa, Micrococcus denitrificans* and other bacteria.

The changes from nitrite to nitrate are more rapid than ammonia to nitrite and, therefore, nitrites are not easily detected in free state in soils. The changes from ammonia to nitrate are even more than ammonification or release of ammonia from organic compounds in soil, and hence there is no large accumulation of ammonia in soil. Only in acid soils rich in organic matter, up to 0.05 per cent of ammoniacal nitrogen may be found. The production of nitrates in soil is influenced by seasonal factors. At low temperature and deficient aeration, there is less nitrification, and as a result the nitrate content may vary. C.M. Hutchinson found that in some Indian soils nitrification can keep pace with ammonification, if the temperature is about 20^{0}C, but at 30^{0}C the ammonification rate exceeded nitrification rate. He attributed this to rapid rise in the H-ion concentration of the soil, which reduced the availability of phosphates to the nitrifying organisms. Hence, he preferred to apply manures like complex organic matter, which encourage slow ammonification.

Denitrification may result in the formation of gaseous nitrogen.

$$RNH_2 + HNO_2 \longrightarrow N_2 + ROH + H_2O$$

$$3HNO_2 \longrightarrow 2NO + HNO_3 + H_2O$$

$$5S + 6KNO_3 + 2H_2O \longrightarrow 3N_2 + K_2SO_4 + 4KHSO_4$$

$$2NO_3^- + 10H \longrightarrow N_2 + 4H_2O + 2OH^-$$

$$2NO_2^- + 6H \longrightarrow N_2 + 2H_2O + 2OH^-$$

$$NO_2 + 2H \longrightarrow N_2 + H_2O$$

There are four important modes of volatilization of nitrogen: (i) non-biological loss, (ii) chemical reaction of nitrate in acidic media to yield nitrogen oxide, (iii) reactions of nitrous acid with ammonium or amino acid, (iv) microbial action to produce N_2 or N_2O. The nitrate formed as an intermediary in the denitrification process may also give rise to either N_2, N_2O or ammonia.

Several soil-dwelling bacteria have the capacity to denitrify nitrate, provided the soil is waterlogged due to poor drainage.

Nitrogen fixation

The soil supports plant growth indefinitely when it is replenished with the nitrogen taken away by the crop plant year after year. The mystery behind this phenomenon was unearthed only a few decades back. The atmosphere contains more than 70 per cent of nitrogen, but only about 0.1 per cent of fixed nitrogen is present in the soil. Small traces of nitrogen from the atmosphere reach the soil in a dissolved state in rain water. Due to electric discharges at the time of lightning, the nitrogen and oxygen in the atmosphere combine to form nitrous and nitric acids.

$$N_2 + O_2 \longrightarrow 2NO$$

$$2NO + O_2 \longrightarrow 2NO_2$$

$$2NO_2 + H_2O \longrightarrow HNO_2 + HNO_3$$

However, the quantity fixed in such manner even under tropical conditions does not exceed 5 to 10 kg of nitrogen per hectare annually. Minute quantities of nitrogen are believed to be fixed in soil due to photochemical action. All these quantities do not account for even a small fraction of nitrogen, which is utilized by the plant. Nitrogenous fertilization of crop plants results in increased yields, somewhat proportional to the dosage of the fertilizer. But in an unfertilized field, the total gain in the nitrogen through plant growth is more than the loss in the nitrogen level in soil. This feature attracted the attention of earlier scientists. The fact that legume plants enriched the soil was known to the Romans. The Greeks and Romans were also aware of the value of leguminous plants as green manure. The Greek philosopher, Theophrastus (370-285 B.C.), in his "Enquiry into Plants" spoke of leguminous plants as 'reinvigorating' the soil. The Roman farmers knew the value of leguminous crops as green manure. Varro de in 37 B.C. said: "legumes should be planted in light soils, not so much for their own crops as for the good they do to subsequent crops". Fuchsius in 1542 was the first to observe the nodules on the roots of leguminous plants. Delechamps in his *Historia Generalis Plantarum* published in the year 1586-87 gave the name *nodules* to the malformations on the roots.

In 1838, J.B. Boussingault conducted some experiments and found evidence that nitrogen enters legume plants by some unknown process, but not the non-legumes. This was questioned by many workers like J.Liebig, and others. In 1858, P.G. Lachmann studied the nodules in the roots of legumes and observed the bacteria. In 1866 M.Woronin further studied the nodule bacteria and found them to be rod-shaped and motile. In 1879, Frank showed that the root nodules are formed due to infection by the organism which he named *Rhizobium*. In 1884, W.O. Atwater conducted some studies with peas and came to the conclusion that the leguminous plants could fix free nitrogen of the air. In the same year, H. Hellriegal and H. Wilfarth brought out the connections between the capacity of legume plants to increase nitrogen content and the bacteria which are present in the nodules of these legumes, thus associating nitrogen fixation with legume nodules. However, the nature of the organism was understood only when M.W. Beijerinck in 1888 isolated the bacterium into pure culture and studied its physiology. The mechanism of root infection by the bacterium was studied by A.Prazmowski in 1889. Later workers like A.I. Virtanen *et al.* showed the differences in the amounts of nitrogen fixed in different plants and strain variations and cross inoculation groups in *Rhizobium*.

As early as in 1879, A.B. Frank described differences in the morphology of the bacterium occurring in different hosts. In 1887, A. Tschirch also reported on similar differences, and Beijerinck in 1888 differentiated X- and Y- shaped forms, rod and pear-shaped forms from three sets of host plants. Later workers substantiated these observations on the morphological differences in the bacterium, which led to the identification of seven different species of *Rhizobium*.

The different groups of rhizobia are generally confined to their own set of host plants, but under certain conditions, they may be promiscuous. In 1939, B.D Wilson, using sterile soil cultures in the laboratory, found far less specificity in the cross-inoculation groups. In 1954, O.N Allen and I.L Baldwin expressed the view that though the lines of demarcation between certain cross-inoculation groups are less well defined, abandoning of the concept hardly solved the problem, and so they recommended that the cross-inoculation groups should be retained. In 1962, G.Rangaswami and G. Oblisami also observed some variations in the cross-infectivity of the bacteria of the cowpea cross- inoculation group, and evidence for a close relationship between cowpea and soybean cross-inoculation groups was indicated in their studies.

Certain rhizobia forming stem nodules in legumes like *Sesbania rostrata* and *Aeschynomene indica* was observed by B.C Dreyfus *et al.* in

1985. The capacity of stem nodules of *S. rostrata* to fix nitrogen is very high, exceeding 150 kg nitrogen fixed per hectare in 52 days. The bacterium has been named as *Azorhizobium caulinodans.* This system offers a unique potential as a green manure in rice cultivation. Rhizobia forming nodules in a non-legume *Parasponia* spp. has also been reported by M.J. Trinick in 1979. The symbiosis between the non-legume *Parasponia* and rhizobia is an extreme example of boundary jumping by rhizobia. Information on the efficiency and quantities of nitrogen fixed by different groups and species of symbiotic microorganisms is rather scattered. Also, wide variations in the quantities of nitrogen fixed by the same *Rhizobium-legume* combination under the influence of different agronomic and environmental factors have been reported.

Physiology of nodule formation

In 1887, Marshall Ward described the appearance of bright spots on the tip or sides of root hair, the curvature of the hair at that point, and the growth of an infection filament from the spot down the root hair. In 1890, A.B. Frank stated that the roots of Leguminosae secreted a substance, which attracted the bacterium. The stimulation of infected cell takes place perhaps due to secretion of some chemicals like indole acetic acid by the bacterium. The rate of movement of the bacteria was studied by some workers. According to some, the host cell wall is dissolved by the action of bacterial enzyme cytase. In 1929, A.I. Virtanen reported that infection occurred through amyloid spots at the tips of very young root hair. As soon as bacteria entered the root hair, an infection thread was formed which grew towards the base, and resulted in the nodule formation because of the division of pericyclic cells or cells adjoining endodermis. At first, the entire nodule tissue is meristematic, but soon the infection thread branches and invades cells in all directions. The infection thread soon liberates bacteria into cytoplasm of the host cell. The bacteria multiply and often fill the entire cell. They are either rod or coccoid and soon become banded and vacuolated. After the nodule formation, vascular bundles are formed, inside which the bacterium establishes contact with host bundles. As the young nodules develop, the vascular bundles are differentiated into nodule cortex and get connected with the vascular system of the host root. As the plants mature, the nodules become brown, the bacteroids disappear, and the nitrogen released for use by the plant. In the final stages, the nodule is cut off from the root by means of a cork layer.

Different types of nodules are formed on the roots, depending upon the host-bacterium relationship. Also, strain variations in the bacterium result in variations in the quantity of nitrogen fixed. Sometimes it results in ineffective strains. Certain differences in the quantities of nodules formed by the effective and ineffective strains of *Rhizobium* have been reported. According to F.J. Bergerson, this was due to ineffective symbiotic relations, as the nodule bacteria are not released from the infection thread; and for the bacterium to become effective, it should be fully developed to a metabolically active bacteroid form within the special host cells in the nodules.

Mechanism of nitrogen fixation

For successful symbiotic nitrogen fixation, a healthy plant growing in sufficient light and an effective nodule forming bacterium are required. The nodule fixes nitrogen only for a short duration when it is in the highest symbiotic relationship with the plant. Nitrogenase is a functional enzyme which reduces nitrogen to ammonia and depends on energy source from ATP. The nitrogenase has two components: one containing Mo-Fe, designated as Mo-Fe protein (nitrogenase), and the other iron protein (nitrogenase reductase). Both the components are essential for nitrogenase activity.

$$N_2 + 3H_2 \xrightarrow{ATP} 2NH_3$$

For this reaction, for every two electron transfer by nitrogenase, four ATP molecules are required as shown in fig. 11.

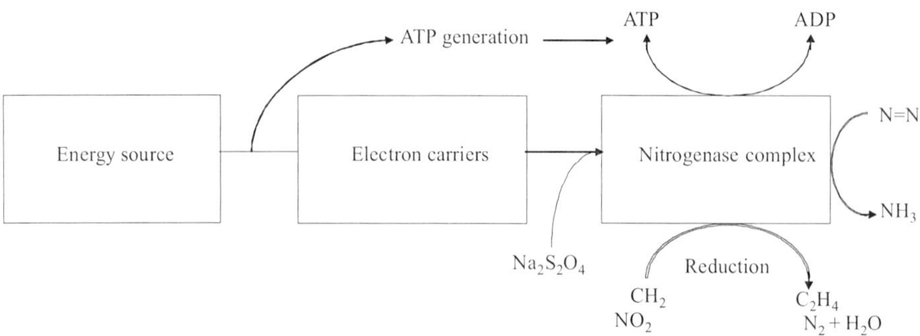

Fig. 11 : Pathway of nitrogen fixation

The energy requirement for biological nitrogen fixation appears to be high, and this becomes the limiting factor in the quantity of N fixed in different legume – *Rhizobium* combinations. Biochemical changes taking place in nodule are illustrated in fig. 12

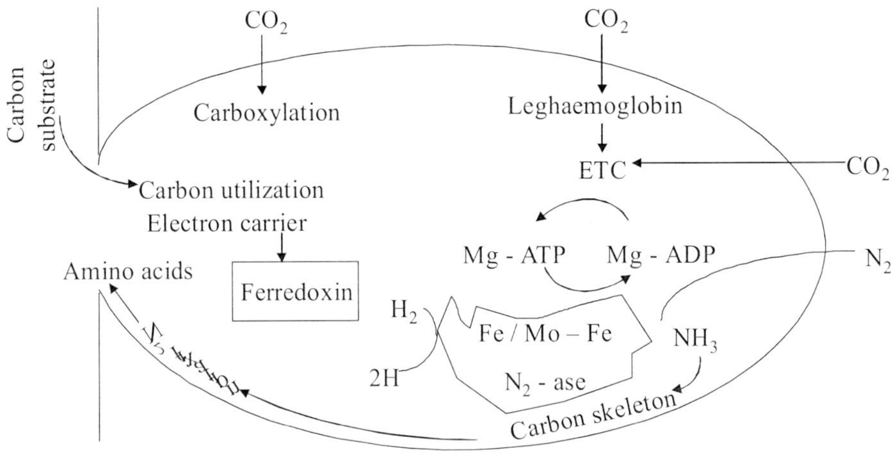

Fig. 12 : Biochemical changes taking place in the nodule

In recent years the genetics and biochemistry of *Rhizobium*-legume nodulation has received considerable attention. Detailed studies on bacterial genetics, plant molecular biology and electron microscopy of bacteria-plant interactions have helped in having a better understanding of the technology of nitrogen fixation. *Rhizobium* bacteria stimulate leguminous plants to develop in a complex series of steps.

Summing up the whole process, it can be said that the rhizosphere organisms convert tryptophan in the root exudates to indole acetic acid (IAA), which brings about the non-specific curling of the root hairs. The capsular polysaccharides of homologous strains of rhizobia play a role in inducing 'shepherd's crook', the site of infection. Pectin methyl esterase produced by rhizobia plus IAA bring about loosening of cell wall material at the shepherd's crook. Due to the production of polygalacturonase by rhizobia, invagination of the root hair takes place, resulting in the formation of infection thread through which rhizobia enter the host. Host nucleus guides the infection thread till it reaches the tetraploid cell, which is the site of developing nodule. Infection thread bursts open to liberate the bacteria in the host cytoplasm. Each bacterium so liberated gets surrounded by a membrane, pinched off from the infection thread. Mitotic divisions of the host cell results in the development of a nodule, and along with it in the distribution of bacteria in cells of the nodule. Bacteria multiply very rapidly, accompanied by enlargement of the membranous envelope surrounding them. Motile rhizobia now enlarge, lose their flagella as also the power of division. They are called bacteroids. The space in between the bacteroids and the membrane gets filled with leghaemoglobin. Nitrogen fixation takes place in these nodules and the effective nodules are pink in colour because of leghaemoglobin. Atmospheric nitrogen gets reduced to ammonia by the nitrogenase system and then to the amino acids in the root nodules which diffuse away and become available to the plant. Legume-*Rhizobium* symbiosis is influenced by a variety of factors like host, bacterial strains, temperature, light, soil pH, phosphorus, combined nitrogen, micronutrients and interaction with other soil organisms.

Rhizobium genetics has thrown light on several gene-controlled steps in the entire symbiotic process. In the initial stages host specificity becomes evident. This is controlled by the nodulation (nod) genes. There may be several nod genes, one or more of them conferring host specificity, the other being common *nod* genes. Some of the *nod* genes induce the host plant to react by producing nodulins. The chemicals have been isolated and characterized as flavanoids, which are three-ring aromatic compounds. These compounds, such as buteolin, naringenin and daidzenin, act as inducers or inhibitors of the various *Rhizobium*-legume systems. Also, some *nod*-genes which may not always be the same as the ones inducing root-hair curling and cell division.

The bacterial infection of the host starts with the complete penetration of the bacteria, which subsequently get released into the host cell. This is accomplished through some bacterial surface components such as heteropolysaccharides, neutral β-glucans, and lipopolysaccharides. These chemicals play the role of signalling the production of osmotic materials necessary for invasion, perhaps to disguise the bacterium during invasion. The production of these bacterial surface compounds is again gene-controlled.

In the plant tissue being invaded by the bacterium, the cell wall and the underlying plant membrane are also involved in invasion. The inside of infection thread is filled with bacteria and with a matrix which is of plant origin. The genes for nitrogen fixation, called *nif* genes, are found in both symbiotic and free-living systems, but in the *Rhizobium*-legume system, the *nif*-genes are distinct. The plant supplies carbon compounds, derived from photosynthesis in the shoot, which helps the bacteria to produce ATP, which is required in large quantities for nitrogen fixation. The leghaemoglobin (Lb) present in the nodule binds oxygen so as to facilitate nitrogen fixation by the bacterium, since presence of oxygen inhibits nitrogenase enzyme of the bacterium. Nodulins are additional nodule specific plant gene products. Both Lb genes and nodulin genes are transcribed in the nodules, whose correct expression depends on the final release of bacteria into target cells. The bacteria in the nodule undergo limited DNA replication and division and then transform into bacteroids. The plant-derived peribacteroid membrane (PBM) forms the boundary or envelop for the bacteroids. There is specific gene-controlled interaction taking place between the bacteroids and the PBM.

The symbiotic activation of *nif*-genes in the rhizobium is dependent on low oxygen concentration, which in turn is regulated by another set of genes called fix-genes, which are common for both symbiotic and free-living nitrogen fixation systems. However, the regulatory pattern may vary in different *Rhizobium*-legume combinations.

The *Rhizobium* bacterial cells and the host cells cooperate intimately in respect of cellular metabolism with the required energy and growth regulation, accompanied by the genetic transcription and translation through DNA. The exchange of carbon sources with that of the nitrogenous substances is balanced as to make the bacterium a symbiont instead of a pathogen, *i.e.* the bacterium, though infective, should ultimately come under the control of the host.

Non symbiotic nitrogen fixation

Nitrogen fixation in soil by free-living microorganisms is known for over a century now. In 1855, J.B. Lawes and J.H. Gilbert found that all plants, except legumes, need ammonia or nitrate as addition to the soil. M. Berthelot in 1888 observed that in unsterilized soil there was a definite increase in nitrogen after some months, but not in sterile soil. In 1894-95 S. Winogradsky was the first to isolate into pure cultures an anaerobic bacterium which could grow in a nitrogen-free medium, and he named it *Clostridium pasteurianum.* In 1901, Beijerinck isolated two aerobic free living nitrogen-fixing bacteria and named them as *Azotobacter chroococcum* and *A.agile.* Inspired by these reports, during the next two or three decades, several workers devoted their attention to the practical applicability of these results, the prevalence of other such bacterial species, and their populations in various soil types. Some of the interesting results obtained by these workers were:

(i) *Azotobacter* spp. were not present in soils below pH 6.0, though some species were isolated from acid soils later.

(ii) Phosphates increased the nitrogen-fixing capacity of *Azotobacter* spp., whereas nitrites inhibited their activity.

(iii) They were either absent or inactive in waterlogged and marshy soils.

(iv) The mechanism of nitrogen fixation and the enzyme systems involved in *Clostridium* and *Azotobacter* seemed different.

Nitrogenase, the key enzyme in nitrogen fixation, has been isolated from several free-living nitrogen-fixing microorganisms, including *Clostridium, Klebsiella, Azotobacter, Chloropseudomonas, Rhodospirillum, Anabaena, Plectonema, Gloeocapsa,* etc. The mutants of *Klebsiella pneumoniae* could also be transferred by conjugation or transduction to non-nitrogen-fixing strains of *K. pneumoniae.* Also, similar transfers could be effected from nitrogen-fixing genera to such non-nitrogen fixers as *Escherichia coli.* The genetic determinants for nitrogen fixation (*nif-genes*) can be separated into those which specify the structure as different from regulation of nitrogenase, and others governing the phenotypic expression of *nif*-genes.

More recent studies on the genetics of free-living nitrogen-fixing bacteria have shown that some of the bacteria such as *Azotobacter vinelandii,* which fix atmospheric nitrogen aerobically without the aid of any specialized structure, contain about 40 times more of DNA than *Escherichia coli.* The *nif*-genes which are present in the bacterium function as the controlling agents of nitrogen fixation. The *ntr*-genes present in the free-living nitrogen-fixing bacteria control the genes that allow growth on poor nitrogen sources. Some of these *ntr*-genes are necessary for transcription of genes involved in utilization of nitrogen, arginine, histidine, proline, etc.

Associative and free-living nitrogen fixation

Azospirillum is of considerable agronomic importance as it is associated with the roots of grasses and is capable of fixing atmospheric nitrogen. Crops grown with pre-treated seed give increased yields up to about 10 to 30 per cent. *Azospirillum* participates in all steps of the nitrogen cycle except nitrification. It can fix atmospheric nitrogen in pure culture and under microaerophilic conditions. A cluster of *nif*-genes has been identified in *Azospirillum,* which are considered to be homologous to those of *Klebsiella pneumoniae. Azospirillum* spp. has been isolated from the rhizosphere of a large number of monocotyledons and a few dicotyledonous plants. In some cases there has been root hair deformation due to the association with the bacterium. The bacterium could invade the cortical and vascular tissues of the host, and lead to enhancement of the number of lateral roots and of root hairs, which helps in increasing the mineral uptake by the plant. Though there is no evidence of host specificity in the *Azospirillum* spp. studied, some affinity in the strains for C_4 as different from C_3 plants has been suggested. Plasmids isolated from *Azospirillum* spp. seem to be associated with nitrogen fixation, but more studies are required to bring out genetics of nitrogen fixation in this group. In the flooded rice fields where microaerobic conditions prevail, some bacteria associated with the rice rhizosphere have been found to help nitrogen fixation. The amount of nitrogen thus fixed has been estimated by the acetylene reduction test at about 42 to 80 kg/hectare per crop season.

The main bacterial genera encountered are *Beijerinckia, Azotobacter, Pseudomonas* and *Flavobacterium.* The tropical soils have much higher *Azotobacter* populations than those found under temperate climates. In soil, their numbers vary from a few to a few hundred per gram of soil. Application of nitrogenous fertilizers drastically reduces the *Azotobacter* population in soil. Inoculation of soil or seed with *Azotobacter* is effective in increasing yields of crops in well-manured soil with high organic matter content as reported by Brown in 1974. Besides the ability to fix atmospheric nitrogen, *Azotobacter* is also known to synthesize biologically active substances such as B-vitamins, IAA and gibberellins in pure cultures. The organism possesses fungistatic properties even on certain pathogenic ones, such as *Alternaria* and *Fusarium.* These attributes of *Azotobacter* explain the observed beneficial effects of the bacterium in improving seed germination and plant growth.

Several experiments conducted in temperate regions show that nitrogen fixation in *Azotobacter* inoculated soils is not more than 10 to 50 kg of N/hectare per year, depending on the availability of carbon sources. Bacterial inoculants containing *Azotobacter* cells under the name 'Azotobacterin' are being produced and used in Russia and East European countries such as Czechoslovakia, Roumania, Poland, Bulgaria and Hungary, where bacterization of seeds with azotobacterin has proved beneficial in increasing yields of crops like wheat, barley, maize, sugar-beet, carrot, cabbage and potato. The increase in the yields of field crops was not more than 12 per cent over corresponding uninoculated controls. Repeated application of *Azotobacter* during different stages of growth of a crop is now being recommended with the object of increasing the number of bacteria in soil. Recent experiments on inoculation of soil with *Azotobacter* along with different doses of inorganic N fertilizer have demonstrated the possibility of saving considerable amounts of N fertilizer while still attaining desired yields of rice. Field experiments with *Azotobacter* in India have revealed that yield increases are not always reproducible. Regarding nitrogen-fixing bacteria, *Beijerinckia* and *Derxia,* occurring in tropical soils, field experiments are needed before any conclusion could be made, to assess their utility and benefits to crop production.

During the last decade, Brazilian workers observed specific and abundant colonization of the rhizosphere of the grass, *Paspalum* with *Azotobacter paspali,* a powerful nitrogen fixer. More recently *Azospirillum lipoferum* has been observed to fix atmospheric nitrogen in the cortical cells of the roots of maize. Substantial increases in yield were reported following the inoculation of sorghum and pearl millet with *Azospirillum brasilense* under several agro-climatic conditions in India. In addition to nitrogen, hormonal effects have been confirmed in Israel. It has also been shown that *Azospirillum* and *Azotobacter,* besides enhancing nitrogen uptake by plants, increase the number of root hairs and root hormone exudation.

When *Azotobacter* is grown in association with cellulose decomposing bacteria, nitrogen fixation by *Azotobacter* is stimulated. *Azotobacter chroococcum* is known to fix more nitrogen, when grown in association with a capsulated bacterium, *Enterobacter aerogenes.* A few fungi and yeasts are reported to fix nitrogen. Algae are reported to fix atmospheric nitrogen, and are important in maintaining the nitrogen balance in rice fields. Blue-green algae are inhibited by ammonium and nitrite in soil and so are not very important in most soils.

Members of the blue-green algae, family Nostocaceae, genera *Nostoc, Anabaena, Aulosira* and *Cylindrospermum,* and a few belonging to the families Rivulariaceae, Stigonemataceae and Scytonemateaceae have been reported to possess the power to fix atmospheric nitrogen by A. Watanabe *et al.* in 1959, R.E. Williams and R.H. Burris in 1952 and R.N. Singh in 1961. They seem to be of significance in rice fields. Singh reported that the algal film that developed on paddy soils in North India belongs mainly to blue-green algae capable of fixing nitrogen, and 40 per cent of nitrogen is excreted by the algae into the substratum, which may become available to rice plants. Watanabe *et al.* showed that species of *Tolypothrix* are active nitrogen-fixers in rice fields in Japan. In desert soils, algae are reported to fix nitrogen in the surface layers.

The blue-green algae abundantly distributed in the tropics play important roles in agriculture. The first and foremost role to be mentioned is nitrogen fixation. Most of the nitrogen-fixing blue-green algae belong to the genera *Anabaena, Anabaenopsis, Aulosira, Cylindrospermum, Nostoc, Calothrix, Scytonema, Tolypothrix, Fischerella, Stigonema,* etc. The amount of nitrogen fixed by the blue-green algae has been reported to vary from 35 to 195 kg/hectare per season. Blue-green algae also add a bulk of organic matter to the soil. Further, they synthesize and secrete several vitamins and growth substances (vitamins B_{12}, auxins and ascorbic acid), which improve, plant growth. Field trials conducted in different parts of India and South East Asia have shown significant increases in grain yields of many rice varieties by inoculation of rice fields with blue-green algae. The farmers in their own fields can continuously prepare algal material for field application. For this purpose, open-air soil culture method is advocated, as it is simple and could be adopted by the farmer himself. A thin layer of soil is spread in galvanized iron trays, 50 cm x 50 cm, filled with 2 cm standing water. The algae are inoculated into the trays and kept in open sun for 5 to 6 days. The algal growth can be dried and stored in polythene bags or used immediately for field inoculation. The production of dried algae on farmers' fields can be a continuous process. The culture can be mixed with lime or in a bucket of water, broadcasted or sprinkled throughout the field, one week before transplanting rice; 10 kg of inoculum per hectare may be needed.

The blue-green alga *Anabaena azollae* is symbiotically associated with the water fern *Azolla.* Azolla is being used as green compost for rice-cultivation in Vietnam, Thailand and China because of its association with the blue-green alga, *A. azollae* which fixes atmospheric nitrogen. An added advantage is that the plant multiplies fast and produces higher yields of green compost (200 to 300 tonnes/hectare per year) than the conventional green manure plants such as *Sesbania, Crotolaria* which are known to yield 30 to 50 tons/hectare per year. It is reported from Vietnam that a tonne of *Azolla* layer enables the rice yield to increase by 10 to 25 per cent over corresponding *Azolla*-free rice field.

Other organisms in nitrogen cycle are the nitrifying bacteria, which oxidize ammonia to nitrite and then nitrite to nitrate. *Nitrosomonas* and *Nitrobacter* are the two bacteria, respectively, responsible for the two-step reaction in soil, and their importance has already been detailed. Martin Alexander in 1978 showed that several heterotrophs belonging to the genera *Pseudomonas, Corynebacterium, Bacillus, Nocardia, Streptomyces, Aspergillus,* etc. produce either nitrite or nitrate from ammonia or other reduced forms of nitrogen. Various factors like aeration, moisture, temperature, pH and organic matter influence the growth and activity of nitrifiers in soil. Denitrification of bound nitrogen to gaseous nitrogen is mediated by numerous species of bacteria. Of these, *Pseudomonas* and *Achromobacter* are the predominant ones in soil.

Sulphur cycle

Sulphur is present in traces in the air, and it makes up about 0.1 per cent of the earth's crust. In the industrial areas, the atmosphere contains high concentrations of sulphur because of the burning of coal. The sulphur in the air may reach the soil through rain water. Most well and river waters contain traces of sulphur. Sulphur and H_2S are abundantly emitted from volcanoes. Sulphur is also present in some springs. Extensive sulphur deposits are found in some parts of the world. Plants utilize sulphur in the dissolved form as sulphate. Its requirement varies with plant species. Generally, cruciferous plants require up to 40 kg of sulphur per hectare, whereas cereals take less than 10 kg per hectare. It may also enter plants through leaf as sulphur dioxide. Inside the plants, sulphur is present as sulphates, parts of amino acids like cystine and methionine, and in some substances like thiamine, biotin and glutathione. Besides, it is known to indirectly influence nodulation in legumes, chlorophyll and other pigment formation, etc. One per cent of the plant material and about 2 per cent of plant protein are composed of sulphur. Animal life also requires sulphur.

The bacteria capable of oxidizing inorganic sulphur compounds could either be aerobic or anaerobic. Their morphology varies from non-filamentous (*Thiobacillus*) to filamentous forms (*Beggiatoa, Thiothrix, Thioploca*). Several fungi and actinomycetes have also been reported to be sulphur-oxidizers (eg. *Aspergillus, Penicillium, Microsporum*). Sulphur-reducing bacteria (*Desulfovibrio desulfuricans*) which reduce inorganic sulphate into hydrogen sulphide have a role in completing the sulphur cycle in nature (Fig. 13).

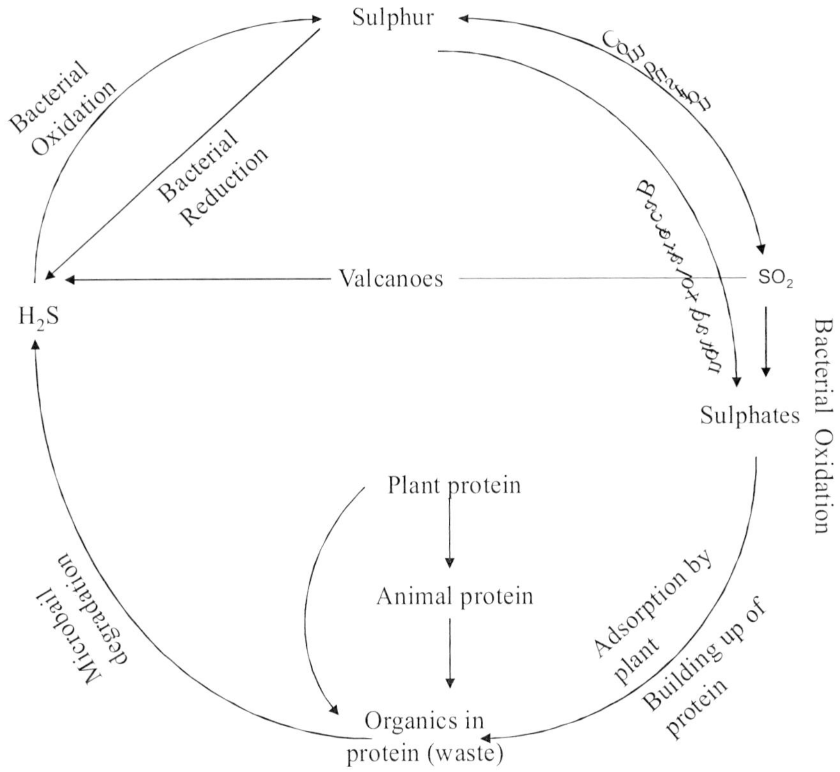

Fig. 13 : An illustration of sulphurcycle

Sulphur enters the soil as organic compounds in plant and animal residues, as elementary sulphur, as H_2S through rain water, or as sulphur, sulphate, etc. in the fertilizers. The complex sulphur containing organic compounds reaching the soil are attacked by several soil-dwellers. The proteins are decomposed to amino acids and amino acids broken down to liberate H_2S which is oxidized to sulphate under aerobic conditions. Under anaerobic conditions, some compounds like mercaptans may be evolved. Under aerobic conditions, the mercaptans are also oxidized to sulphates. Hydrogen sulphide, when washed into soil, is utilized by the characteristic autotrophs known as sulphur bacteria, of the genus *Thiobacillus* spp. *T. thioparus, T. thiooxidans* and *T. denitrificans* are strict chemosynthetic autotrophs, utilizing the energy liberated in the oxidation of sulphur.

$$S_2 + O_2 + 2H_2O \longrightarrow 2H_2SO_4$$

$$2H_2S + O_2 \longrightarrow S_2 + 2H_2O$$

Thus, sulphur and H_2S are oxidized to yield sulphuric acid, which can react with calcium phosphate and other substances. Sometimes the sulphate is reduced to sulphite by the sulphate-reducing bacteria like *Microspira desulphuricans* and *Sporovibrio desulphuricans.* The reduction process is particularly favoured by the alkaline and anaerobic conditions of the soil, as the organisms use sulphate as a source of oxygen supply. In this process, H_2S is released, which gives the characteristic odour. *Beggiatoa alba* is filamentous and is chemosynthetically autotropic. It deposits sulphur, utilizing H_2S:

$$H_2S + \frac{1}{2}O_2 \longrightarrow S + H_2O \text{ (41,400 cal of energy)}$$

$$S + H_2 + 1\frac{1}{2}O_2 \longrightarrow H_2SO_4 \text{ (118,000 cal of energy)}$$

Besides the strict autotrophs of the genus *Thiobacillus,* there are several other members which are facultative autotrophs capable of reducing H_2S and derive energy from it. They belong to the family *Achromobacteraceae, Beggiatoaceae, Chlorobacteriaceae, Thiobacteriaceae* and *Thiorhodaceae.* These organisms utilize H_2S as a hydrogen donor for the reduction of CO_2. They are found in sulphur springs, sewage, stagnant deep waters, etc. They may also be present in water pipes and cause serious offensive odour and obstruction to the flow. In the presence of iron, for example, in water pipes, the bacteria may corrode the pipes.

There are also some photosynthetic autotrophic bacteria falling under *Thiorhodaceae,* which are purple to red. They contain sulphur granules and can oxidize various sulphur-containing compounds to sulphate and at the same time photosynthetically reduce CO_2.

$$H_2S + 2CO_2 + H_2O \xrightarrow{\text{light}} H_2SO_4 + 2(CH_2O)$$

In the case of green bacteria of Chlorobacteriaceae, H_2S serves as the hydrogen donor.

$$CO_2 + 2H_2S \xrightarrow{\text{light}} (CH_2O) + H_2O + 2S$$

Sulphur containing amino acids such as cysteine are enzymatically attacked by many heterotrophic bacteria to produce H_2S.

$$\underset{\text{cysteine}}{CH_2.SH.CHNH_2.COOH} + H_2 \longrightarrow \underset{\text{pyruvic acid}}{CH_3.CO.COOH} + H_2S + NH_3$$

Sulphates such as calcium sulphate can be attacked under anaerobic conditions by members of the *genus Desulfovibrio* to release H_2S:

$$CaSO_4 + 4H_2 \longrightarrow Ca(OH)_2 + H_2S + 2H_2O$$

The H_2S may be photosynthetically utilized and elemental sulphur may be released.

The sulphates and sulphuric acid, when dissolved in water, are made available for plant growth. The plants utilize the sulphates to form various amino acids, hormones, growth factors, etc. They are either eaten away by animals or returned to soil as organic waste. The plant parts eaten by the animals are also in some form or other returned to the soil. When the various complex organic sulphur compounds reach the soil, they are attacked by the soil organisms and the cycle of events continues.

Phosphorus cycle

Phosphate constitutes nearly 0.1 per cent of the earth's crust. Rock phosphates are present in parent rocks. They occur in soil in inorganic and organic forms. The inorganic forms are derived from parent rocks or through fertilizer application and manuring with bone meal, and are soluble in water when present as phosphates of Na, K, Ca, Mg, etc. In cultivated soil they are present in abundance-about 1,100 kg / ha, but most of them are not available to plants; only about 1 per cent of the total soil phosphorus is in available form. Therefore, when phosphates are applied as fertilizers, they are utilized more by the plants than the soil phosphates. The common organic forms of phosphorus are in lecithin, phytin, nucleic acids, nucleoproteins, etc. They are found in the bones of man and animals as calcium phosphate, and as nucleic acid and nucleoproteins in the cytoplasm of living cells of all nature. The residues of animal, man, plants, birds, etc. contain several phosphates. When they reach the soil, they are acted upon by several microorganisms. The organisms break down the phosphorus-containing compounds, with the liberation of mineral elements, such as Ca, Fe, Na, and this process is known as *mineralisation.* Sometimes, the bacteria remove soluble phosphates in soil and use them for cell synthesis. On the death of the bacteria, the phosphate is made available for plants. A fraction of the phosphate is also in soil due to leaching. Inoculation of organic compost with phosphate solubilizing *Bacillus polymyxa* and nitrogen-fixing *Azotobacter chroococcum* help in improving the quality of the manure by reducing the C:N ratio from 15 to 12 and substantially improving the available phosphate.

The activity of microorganisms in phosphate solubilization is substantially influenced by various soil factors such as pH, moisture and aeration. Many fungi and bacteria (e.g.*Aspergillus, Penicillium, Bacillus*) are potential solubilizers of bound phosphates, as revealed by experiments *in vitro*. These organisms produce organic acids like citric, glutamic, succinic, lactic, oxalic, maleic, fumaric, tartaric and α-ketoglutaric acids, which are responsible for solubilization of insoluble forms of phosphorus. These organisms have a role in completing the phosphorus cycle in nature (Fig. 14).

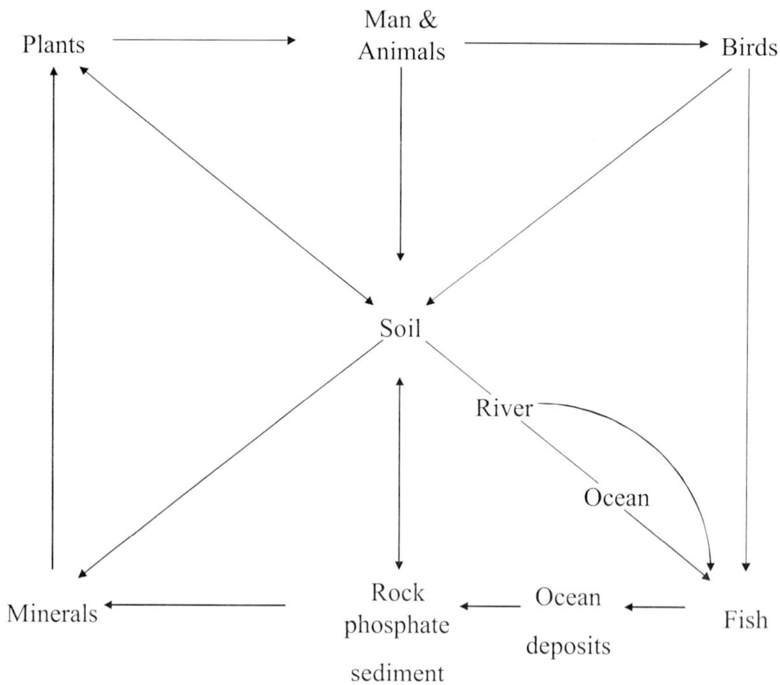

Fig. 14 : An illustration of phosphorus cycle

Other nutrients

Besides O, H, N, P and S discussed already, plants require K, Ca, Mg, Fe, Mn, Zn, Cu, Co, B and Cl, for their growth. It is quite obvious that soil microorganisms play important roles in the release of some of these elements in soil. The availability of Zn, Cu, Co, B, Mo, etc. in soil has been studied more accurately by microbiological assay using nutritionally deficient strains of *Aspergillus niger* and some bacteria like *Azotobacter chroococcum* and algae like *Anabaena cylindrica* and *Chlorella* spp. It is believed that oxidizing pyrite (FeS_2) in the coal and rock strata gives rise to ferrous sulphate, and the sulphur oxidizing bacteria change ferrous to ferric sulphate. There is evidence to show that chemosynthetic bacteria are associated with bituminous coal mines. *Thiobacillus ferrooxidans* is a strictly autotropic iron bacterium. Some soil microorganisms utilize ferrous salts as energy source. By oxidizing ferrous sulphate to ferric compound, energy is liberated, which is utilized by the organisms to assimilate CO_2 from carbohydrates.

$$2FeSO_4 + 3H_2O + 2CaCO_3 + O \longrightarrow 2Fe(OH)_3 + 2CaSO_4 + 2CO_2$$

$$4FeCO_3 + 6H_2O + O_2 \longrightarrow 4Fe(OH)_3 + 4CO_2$$

It is believed that the iron bacteria are responsible for the iron deposits in some parts of the world. They grow normally at a low temperature range of 0 to 20°C and are found commonly in cool springs and brooks which are rich in iron salts. These organisms are capable of taking up iron and accumulating or depositing it on its own cell surface in the form of ferric hydroxide. They can be broadly divided into autotrophic iron bacteria and non-autotrophic forms. The autotrophic forms utilize CO_2 as carbon source by deriving energy by the oxidation of ferrous to ferric salts. The other types of iron bacteria are capable of utilizing both inorganic and organic substances and derive energy, and at the same time oxidize ferrous salts. Here again, there is decomposition of iron salts on the cell surface. *Gallionella, Spherotilus, Clonothrix, Leptothrix, Tolypothrix and Crenothrix* are important genera of this type of iron bacteria. They grow in water rich in iron salts or public water pipes which are old and corroded. *Spherotilus natans* and *S. discophora* are important species falling under this category. They form yellowish or brown or red flocculent masses. Many algae, including blue-green, are capable of reducing ferric iron to form ferrous sulphate in water.

Potassium is found in soil in both organic and inorganic forms. In the plant residues and manure, they are in a concentration of 0.2 to 2.0 per cent, whereas in microbial cells they vary greatly. When microbial activity is rapid, the organic acids formed help in releasing potassium from the inorganic salts in zeolite. They are taken up by the fungi and bacteria in soil and are stored in their cells. Calcium and magnesium are also released from inorganic and organic forms in soil by microorganisms and are later released for plant growth.

It is now well-known that soil microorganisms excel other biological systems in synthesizing certain chemical substances. Their synthetic powers are complicated and the structures of the chemical molecules synthesized by them are intricate. Several such chemicals have been isolated and studied, which include antibiotics, hormones, vitamins, alkaloids, steroids, quinones, etc., and in many cases several mineral elements like Fe, S, Cl, Mg, Co are incorporated. These substances are derived by the organisms from simple inorganic salts or from complex organic substances. In most of these cases, the chemicals, on their release from the microorganisms in this cycle of events is very important and indispensable. The elements like Mo, Cu, Fe, Zn, Co, B and others are important for the growth of some organisms. These substances reach the soil from different sources and are later made available for the plants through microbial and other physicochemical processes. Several mineral compounds like silica are rendered soluble by the action of soil microorganisms like diatoms for their growth, and is also utilized by higher plants. In some cases the microorganisms change the inorganic compounds without taking them into their cell. These processes are mainly the means of oxidation or reduction. Typical examples are reduction of nitrate to nitrite, sulphate to sulphide, and ferric to ferrous substances.

Several toxic substances, both organic and inorganic, may reach the soil. They may affect the microbial population, if not acted upon. There are some bacteria and fungi which can attack these substances and convert them into non-toxic forms. Arsenic compounds can be attacked by some species of *Penicillium* and *Aspergillus*, and free arsenic may be released into the atmosphere. Similarly, selenium present in organic tissues may be volatalized by some fungi and bacteria. Some herbicides and insecticides which reach the soil are toxic to soil microorganisms. But they are attacked by a few soil bacteria to break down the molecule into simple non-toxic substances. Thus 2, 4-dichlorophenoxyacetic acid, 2, 4-dinitroorthocresol, etc. are reported to be decomposed slowly by soil microorganisms.

Soil-Plant-Microbe Interactions

Soil is living system with dynamic physico-chemical and biological properties. Over the past billion years or so, numerous biological species have evolved on the soil and earth. The evolutionary processes have brought incomplete interrelationship involving non-living and living systems. There exists certain specific relationships between soil, plant and microorganisms, analogous to man-plant-animal systems, though at much different physical dimensions. However, our knowledge of the interrelationships is very limited as compared to the knowledge yet to be gained. The more we probe into relationships, the more we are perplexed about their complexities. An attempt is made here to briefly bring out the interrelationships.

Soil and plant

The soil in which plants grow varies widely in its physical, chemical and biological properties. Several of these factors influence plant growth, both directly and indirectly. Man further interferes with such natural influences, directly through various agricultural practices and indirectly through his socio-economic activities. Soil, being loaded with millions of microorganisms, influences the plant. Besides, the air and water-borne microorganisms also affect the plant and to a limited extent the soil. As a result, in a given location, the interactions between the soil, plant and microorganisms become distinct and dynamic. The innumerable factors acting on the interrelationships lend dynamism and uniqueness under a given set of environmental conditions. Besides supporting the growth of various biological systems, soil also serves as a medium in which the organic residues in the form of plant and animal wastes are disintegrated and decomposed. Not only the large volumes of such wastes are deformed and converted into harmless or inert substances, but also several pathogenic forms of microbes are suppressed and eliminated when incorporated in the soil. Thus, soil functions as a silent scavenger for the entire humanity, but for which this planet would become unfit for habitation.

The plant species which are supported by soil vary in their size and other morphological features. They have evolved themselves into innumerable genetic species and varieties. They may be annuals or perennials and dwarf plants or gigantic trees. Their height, girth, foliar spread and physiological conditions differ widely, and these in turn influence the growth and morphology of their root system. The growing plant, on the other hand, exerts some influence on the soil and its properties. Not only the physical structure is improved through crop cultivation, but also the fertility level improves through addition of the crop residues, resulting in increased microbial activity. This would vary with plant species and soil type. An intricate relationship between the plant root and soil systems develops and functions in a dynamic equilibrium in a given location and biosphere. Any change in the growth pattern of the plant or the soil qualities alters the equilibrium. Over the centuries man has learnt to influence the growth of plant, both directly through soil treatments and cultural practices and indirectly through environmental and ecological changes. Thus, different intensities of interrelationships between plant and soil have developed upon whether it is an annual or perennial plant, wild or cultivated species, or cultivated under intensive monocropping, or mixed and multiple cropping systems.

Microbes and plant growth

The third dimension to the complex relationship is added by the existence of microorganisms in nature. The microorganisms were the first to appear on this planet, and through evolutionary processes, a large number of morphological and physiological forms have developed. Being minute and microscopic, they are universally present in soil, water and air in large number per unit volume of space. By their number and qualities, and by their universal presence they influence all animate and inanimate objects on earth. In respect of soil-plant relationships also, the microorganisms influence them both directly and indirectly. The soil microorganisms help the plant to obtain nutrients from the soil by means of their intimate relationships with the root system. The efficiency of the plant in absorbing the nutrients from the soil is also largely dependent on the associated microorganisms. Besides, the microorganisms present in the air and water, when brought into contact with plant, influence its growth and functions. Such influences through soil, water and air may mostly be beneficial, but at times they may lead to harmful or adverse effects on plant growth. In recent years, the role of microorganisms in influencing plant growth is being increasingly understood.

Seed-borne microflora and spermosphere effect

As early as in 1904, M. Duggeli reported that healthy seeds carry specific bacterial flora in respect of number and species. Since then, innumerable reports have appeared in the literature on the quantity and quality of microorganisms carried, by the seeds of different plant species both externally and internally. While most of the microorganisms are harmless, some may be positively beneficial and a few others may be pathogenic under certain favourable conditions.

The role of seed-borne microflora in seed germination and subsequent growth of the seedling has been investigated by some workers. It has been shown that some organisms have beneficial effects on the germinating seed through some biological products, such as growth hormones. It has been reported that the germinating seed excretes some chemicals, which influence the quality and quantity of the microorganisms on the seed and in the immediate vicinity. In 1959, G. Picci defined the region of such influence around the seed as *spermosphere,* and the phenomenon as *spermosphere effect.* When the seed is sown in soil, certain interactions between the seed-borne microflora and the soil microorganisms take place, and these too, under the influence of the chemicals exuded by the germinating seed. Such interactions vary with the seed and its microflora, the soil and its microflora, as also the environmental conditions under which they interact. When the seed is pre-treated with a fungicide or with a biological agent, the interaction is largely interfered with. Not only the microflora of the spermosphere is altered, but also the quality and quantity of the microorganisms moving over to the root surface of the emerging seedling get altered. Thus our interest in spermosphere effect is more than academic since we can obtain immediate benefits through practical utilization of the basic knowledge on the artificial manipulations of the microorganisms in the root region through seed.

The influence of germinating seed on the microflora may be general or specific, more often specific to one or more groups of organisms. Thus, when a seed carrying the natural and artificially pre-treated load of microorganisms is sown in soil, certain microorganisms are selectively activated and others suppressed. The organisms loaded through pre-treatment of the seed by virtue of numbers often get more support and become dominant. This is why *Rhizobium, Azotobacter, Azospirillum,* etc. applied through pre-treatment of seed, get established on the root surface of the germinating seed and benefit the plant. While the interactions between the microflora and the germinating seed continue in the spermosphere region, the teeming millions of microorganisms present in the soil also get activated, almost instantaneously. The soil organisms compete with seed-borne organisms, both for space and nutrients. The physiological quantities of the seed-borne organisms as compared to the soil organisms, and the quantity and quality of the chemicals exuded by the germinating seed, decide the final shape of things in the region. Subsequently, the microorganisms in the spermosphere region move to the rhizosphere. It has been reported that selected microorganisms from the seed surface could move through the spermosphere to the rhizopshere within a period of three days. This would, however, vary with the plant species, the nature and population of the microorganisms concerned on the seed, the soil qualities, particularly the nature of its microflora, and the environmental conditions for germination and growth of the seed in the soil.

It has been further demonstrated by some authors that various methods of chemical pre-treatment of the seed, such as coating with organo-mercurials, bring about marked changes on the rhizosphere microflora of the seedlings, thereby clearly indicating the possibilities of altering the plant-root microorganisms relationships in the soil through the seed. Also, when the seed is infected internally or infested externally with a pathogenic organism, in sufficient population strength, the competition for space and nutrients arises between the pathogen and the other seed-borne microorganisms. This is followed by a competition between the resultant spermosphere organisms and the soil microflora in the rhizosphere region. When such a seed is pre-treated with an effective plant protection chemical, the competition in the spermosphere is either partially or fully eliminated, and hence the diseases are checked. When the seed is pre-treated with organic matter, such as cowdung slurry or with certain antagonistic or rapidly propagating saprophytic microorganisms, competition for space and nutrients between the pathogen and the non-pathogens sets in and depending on the efficiency of either group, one is suppressed and eventually get eliminated. Internally and externally seed-borne pathogen of cotton, *Xanthomonas campestris* pv. *malvacearum* is controlled when the seed is pre-treated with cowdung slurry containing loads of saprophytic organisms, and sown in the soil. Also, when a seed infected with a pathogen is sown in unsterile as compared to sterile soil, the spermosphere effect is intensive enough to suppress the pathogen in the former, whereas in the latter case the pathogen becomes highly virulent, unchecked by other organisms.

Rhizosphere effect

When plant grows in soil, in the initial stages the spermosphere microorganisms may dominate in the rhizosphere of young roots. As the plant continues to grow, to a large extent, soil microorganisms take over, though in many cases, the seed and spermosphere microorganisms may also persist. However, not all groups of microorganisms present in the soil reach the rhizosphere; only selected groups capable of competing with the others and establishing a beneficial relationship with the root system finally multiply and become predominant. In numbers, the rhizosphere microbes are ten to hundred times more than those of the soil. Their composition would, however, depend upon the quality and quantity of the nutrients in the rhizosphere. The overall metabolic activity of the microorganisms in the rhizosphere is several times more than in the soil farther away from the influence of root zone.

While the nutrients present may be primarily of plant origin, and may vary with the plant species and varieties and physiological conditions of the plant, there may also be substances released by the early inhabitants of the rhizosphere among the microorganisms, which become available for the others. The spectrum of chemicals exuded by the plant roots varies widely, and hence their influence on the microflora also varies widely. Another source of nutrients for the microorganisms in the region is the sloughed off root epidermis which gets disintegrated through microbial action. Some of the substances released by the microorganisms in the region get absorbed by the plant root, the organisms playing an active role in mobilizing the immobilized plant nutrients. There are also evidences to show that the microflora of the surface layer of the soil are carried to the deeper layers by the growing root system. How long such 'displaced' microorganisms persist in the deeper layers depends on various soil environmental factors. However, the 'ploughing in' of the organisms improves the fertility level of the soil. Thus, a dynamic equilibrium prevails in the rhizosphere in the respect of microbial population and availability of plant and microbial nutrients. The complex system of biological and biochemical relationship in the rhizosphere continuously changes because of the changing physiological condition of the plant.

It has been established that through foliar chemical sprays the exudation pattern of the roots could be altered. When ^{14}C and ^{32}P chemicals were sprayed on plant foliage, they were detected in the root exudates within 12 to 36 hours. Such changes in the root exudation within 12 to 36 hours influence the rhizosphere microflora. Foliar applications of organic substances, including plant protection chemicals, growth regulators and plant nutrients, have been found to cause marked changes in the number and qualities of microorganisms in several cereals and leguminous plants. If we can understand the intricate relationships between plant roots and the rhizosphere microorganisms under the direct influence of soil amendments and foliar sprays, we will be able to manipulate in such a manner as to encourage such organisms which excrete growth promoting auxins and mobilize plant nutrients from the soil. This would benefit plant growth, encourage inhibition of plant pathogenic organisms in the vicinity, and enrich the soil with added microbial biomass.

The presence of plant pathogen in the soil and/or in the rhizosphere further complicates the interrelationship. The balance between the rhizosphere microflora and plant pathogen, *vis-à-vis* the soil microflora and the plant pathogen are important in host-pathogenic relationships. Unless the plant pathogen is capable of competing with other soil organisms and penetrating the barrier of a mantle of a wide variety and large number of rhizosphere microorganisms, it cannot reach the root surface and cause infection. In this context, the biochemical quantities of root exudates and the presence of antagonistic microorganisms, if any, have a role to play in encouraging or inhibiting the pathogen in the region. In effect, several mutualistic, commensalistic, competitive and antagonistic interrelationships prevail in the rhizosphere. There are reports, which show that the quality and quantity of microorganisms present in the rhizosphere of disease-resistant crop varieties are significantly different from those of susceptible varieties. In some cases organisms specifically antagonistic to the concerned pathogen have been isolated from the rhizospheres of resistant crop varieties, but not from the susceptible ones. In a few studies, attempts to establish a pathogen in the rhizosphere of a non-host or resistant host varieties were not successful, whereas such attempts could be easily done in susceptible varieties. It has also been demonstrated that through artificial means, such as fertilizer application and organic amendments of soil, foliar chemical sprays, the number and qualities of the group of antagonistic microorganisms in the rhizosphere could be increased.

It has already been mentioned that changes in the physiological conditions of the plant influence the rhizosphere microflora through root exudations. Hence, significant changes in the rhizosphere microflora of diseased plants have been reported. Such changes depend on the nature and severity of the disease. The effects due to systemic diseases are bound to be different from those of localized leaf spots. Systemic virus diseases cause marked changes in the plant morphology and physiology to drastically alter the rhizosphere microorganisms. In this context, mention may be made on the reports on excretion of hydrocyanic acid by the roots of sorghum (which suppresses certain microorganisms in the root zone) and on the presence of phosphatic enzymes in the rhizosphere region of leguminous plants due to selective activation of phosphobacteria, resulting in increased availability of phosphate to the plants. Also, in certain plants, specific plant pathogens are stimulated either directly through root exudates or indirectly under the influence of metabolites of microbes, which become available for the pathogens in the rhizosphere. There are some reports on the selective stimulation of *Fusarium,* root infesting nematodes, and *Pseudomonas* in the rhizospheres of the respective susceptible hosts. It is worthwhile to understand the influence of root exudations on rhizosphere microflora, and the ways and means of selectively activating them for the benefit of plant growth.

There is growing interest in respect of allelopathy in agriculture, whereby the growth of one plant species is suppressed by another. This phenomenon is seen when some crop plants suppress the weeds in the field. The main mechanism of action is believed to be through allelo-chemicals released by the roots of the suppressive plant species. When the concentration of such chemicals in the weed species is relatively high to suppress the crop plant species, the weeds persist for long in the soil. Such chemicals, when excreted by the weed or crop plants, both directly and indirectly influence the microorganisms in spermosphere, rhizosphere, rhizoplane, etc. which in turn influence the germination of seed and growth of plants. In mixed and relay cropping as well as crop rotational systems, the allelopathic interactions get either accentuated or reduced through the interference of the soil microflora. If we could understand the biochemistry and physiology of allelopathy, we could develop effective technologies for controlling the weeds in crop fields.

Phyllosphere effect

J. Ruinen in 1961 showed the importance of external leaf surface as an environment for microorganisms. Since then S. Sinha and his associates and several other workers have reported on the interrelationship between the plant foliage and the microorganisms found on the surface. The quantity and quality of microorganisms in the region, named as *phyllosphere,* vary with the plant species, and its morphological, physiological and environmental factors. The age of the plant, its leaf spread, morphology and maturity level, and the atmospheric factors under which it grows, greatly influence the phyllosphere microflora. Both native or resident organisms and casual depositors are found on the leaf surface. More organisms are retained on a surface which are hairy, waxy or highly cuticulate than on a smooth dry surface. Leaf diffusates and exudates have been reported to encourage growth and multiplication of the phyllosphere microorganisms. Since large variety of microbes, both pathogens and saprophytes, are found on the leaf surface, the stimulation of growth in the microorganisms varies widely, and is often selective. It is claimed that such selective stimulation brings in interactions, including antagonistic effects, among the organisms. The metabolites of the microbes in turn affect the plant growth. There are reports of nitrogen fixation by certain bacteria prevalent in the phyllosphere. Some studies have shown that the entry of plant pathogens into the leaf is facilitated by the physiological activities of non-pathogens. Thus the phyllosphere effect seems to be specific to a given set of plant foliage and its microbial load. There are also reports on the suppression of phyllosphere microflora due to environmental pollutions caused especially by cement and fertilizer industries. Any change in the phyllosphere effect affects plant growth, which in turn affects the physiological activity of the root system. Such changes in root result in an altered pH and spectrum of chemical exudations, causing a change in the rhizosphere microflora. Thus, there exists a link between phyllosphere microflora and rhizosphere microflora.

Chemical sprays on the foliage interfere with the phyllosphere microbes. Plant protection chemical sprays or dusts may destroy most of the microbes, but not all of them. Plant nutrients, when sprayed on the foliage, cause both direct and indirect effects on the phyllosphere microflora. While the plant nutrients might also be nutrients for the microbes, there would be differential utilization of the nutrients by different groups of microorganisms on the leaf. When the nutrients are absorbed by the foliage and utilized for plant growth, the phyllosphere activity of the plant gets altered, and brings about changes in the leaf exudation pattern. This has an indirect influence on the microbial activity in the phyllosphere. If a plant hormone is sprayed on the foliage, it may or may not have any direct influence on the leaf surface microflora, but it brings about significant changes in the physiology of the plant. This in turn alters the diffusates and exudates from the leaf and consequently the phyllosphere microflora. Altering the physical structure of the soil through cultivation practices helps in improving the root structure in a given plant species, which in turn alters the rhizosphere microflora. Adoption of different practices of crop rotations, *i.e.* mixed cropping and multiple cropping, in a given soil activates root exudates in various layers of the soil in different ways, which helps in enhancing the rhizosphere effect. Often such effects have been found to be helpful in suppressing soil plant pathogens.

Though complex, the rhizosphere technology is a very valuable tool in the hands of agricultural scientists to manipulate and influence the soil-plant-microbe interrelationships to give more benefit to the farmers. The field is wide open and rich for exploration.

References

Alexander, M. 1977. *Introduction to Soil Microbiology* (2nd edn.), John Wiley & Sons, New York, USA.

Angle, J.S. 1994. Viruses. In: *Methods of Soil Analysis, Part 2: Microbiological and Biochemical Properties.* (eds. R.W. Weaver *et al.*). Soil Science Society of America Pub., Madison, USA. pp. 107-118.

Atlas, R.M. and Bartha, R. 1993. *Microbial Ecology: Fundamentals and Applications.* (3rd edn.) Benjamin/Cummings, Redwood City, California, USA.

Black, C.A. 1968. *Soil Plant Relationship.* John Wiley & Sons, London, UK.

Bowen, G.D. and Rovira, A.D. 1976. Microbial colonization of plant roots. *Annual Review of Phytopathology.*, 14: 121-144

Brady, N.C. and Weil, R.R. 1996. *The Nature and Properties of Soils* (11th edn.), Prentice-Hall, New York, USA.

Coyne, M.S. 1999. *Soil Microbiology: An Exploratory Approach.* Delmar Publishers, New york, USA.

Curl, E.A and Truelove, B. 1986. *The Rhizosphere.* Springer-Verlag, New York, USA.

Dar, G.H. 2000. *Soil Microbiology and Biochemistry.* NIPA Pub., New Delhi, India.

Dickson, C.H. and Preece, T.F. (eds.) 1976. *Microbiology of Aerial Plant Surfaces.* Academic press, London, UK.

Edwards, C.A., Stinner, B.R., Stinner, D. and Rabatin, S. 1988. *Biological Interactions in Soil.* Elsevier Publication, Amsterdam, Holland.

Foth, H.D. 1990. *Fundamentals of Soil Sciences.* (8th edn.) John Wiley and Sons, New York, USA.

Gray, T.R.G. and Williams, S.T. 1971. *Soil Microorganisms.* Oliver & Boyd, Edinburgh, UK.

Harris, R.F. 1981. Effect of water potential on microbial growth and activity.In: *Water Potential Relations in Soil Microbiology,* (eds. J. F. Parr *et al.*) Soil Science Society of America, Madison, Wisconsin, USA. pp. 23-95.

Ingraham, R.E. 1994. Nematodes. In: *Methods of Soil Analysis, Part 2: Microbiological and Biochemical Properties* (eds. R.W. Weaver *et al.*) Soil Science Society of America, Madison, Wisconsin, USA. pp. 459-490.

Killham, K. 1994. *Soil Ecology*. Cambridge Univ. Press, Cambridge, UK.

Kleister, D.L. and Cregan, P.B. (eds.) 1991. *The Rhizosphere and Plant Growth*. Kluwer Academic, Dordrecht, Germany.

Lynch, J.M. 1983. *Soil Biotechnology: Microbiological Factors in Crop Productivity*. Blackwell Scientific Publications, Oxford, UK.

Madigan, M.T., Martinko, J.M., Dunlap, P.V. and Clark, D.P. 2009. *Brock Biology of Microorganisms* (12th edn.) Pearson Benjamin Cummings, New York, USA.

Metting, F.B. (ed.). 1993. *Soil Microbial Ecology*. Marcel Dekker, New York, USA.

Parkinson, D. 1994. Filamentous fungi. In: *Methods of Soil Analysis, Part 2: Microbiological and Biochemical properties*. (eds. R. W. Weaver *et al.*) Soil Science Society of America, Madison, Wisconsin, USA. pp.329-350.

Paul, E.A. and Clark, F.E. 1989. *Soil Microbiology and Biochemistry*. Academic Press, New York, USA.

Rangaswami, G. 1986. Soil-plant-microbe interrelatioships. *Indian Phytopathology.*, 41:165-172.

Richards, B.N. 1987. *The Microbiology of Terrestrial ecosystems*. Longman, Harlow, England.

Russel, S.R. 1977. *Plant Root System – Their Function and Interaction with the Soil*. McGraw-Hill, London, UK.

Ryder, M.H., Stevens, P.M. and Bowen, G.D. (eds.) 1994. *Plant Productivity with Rhizosphere Bacteria*. Division of Soils, CSIRO Pub., Glen Osmond, South Australia.

Spurr, H.W. 1990. The phyllosphere. In: *New Directions in Biological Control: Alternatives for suppressing Agricultural Pests and Diseases* (eds. R.R. Baker and P.E. Dunn). Alan R. Liss Inc., New York, USA. pp. 271-278.

Stevenson, F.J. 1982. Origin and distribution of nitrogen in soil. In: *Nitrogen in Agricultural Soil*, (ed. F.J. Stevenson), Agronomy Society of America, Madison, Wisconsin, USA. pp. 1-42.

Subba Rao, N.S. 1982. *Advances in Agricultural Microbiology*. Oxford & IBH Publishing Co., New Delhi, India.

Subba Rao, N.S. and Dommergues, Y.R. (eds.) 1998. *Microbial Interaction in Agriculture and Forestry. Vol. 1.* Oxford & IBH Publications Co., New Delhi, India.

Tate, R.L. 2000. *Soil Microbiology.* John Wiley & Sons, New York, USA.

Tilak, K.V.B.R and Ranganayaki, N. 2005. Associative effects of beneficial rhizosphere microorganisms for sustainable agriculture. In: *Frontiers in Plant Sciences* (eds. K.G. Mukerji, K.V.B.R. Tilak, S.M. Reddy, L.V. Gangawane, P. Prakash and I.K. Kunwar), I.K. International Pvt. Ltd., New Delhi, India, pp. 143-161.

Waksman, S.A. 1957. *Soil Microbiology.* John Wiley & Sons, New York, USA.

Wellington, E.M.H. and Toth, I.K. 1994. Actinomycetes. In: *Methods of Soil Analysis, Part 2: Microbiological and Biochemical Properties,* (eds. R. W. Weaver *et al.*) Soil Science Society of America, Madison, Wisconsin, USA. pp. 260-90

Wood, M. 1989. *Soil Biology.* Blackie, Glasgow, UK.

Chapter-3

Nitrogen Fixing Microorganisms

Nitrogen is the most important element on which depends growth and productivity of plants. It is also the most abundant molecule in the atmosphere (80%), yet it cannot be utilized by plants. The plant utilizable form of nitrogen is either, its most oxidized form, the nitrates or the most reduced state, the ammonia. The indiscriminate use of nitrogenous fertilizers amounts to depletion of non-renewable fossil fuels used in fertilizer production. Sustainable agriculture depends largely on renewable resources and on-farm nitrogen contributions are achieved largely through biological nitrogen fixation (BNF), which helps in maintaining and/or improving soil fertility by using atmospheric nitrogen. Annually BNF is estimated to be around 175 million tonnes of which close to 79% is accounted for by terrestrial fixation. This illustrates the importance of BNF in the context of global N cycle. The BNF offers an economically attractive and ecologically sound means of reducing external N inputs and improving the quality and quantity of internal resources. Legumes have been an important component of agriculture since ancient times. It is widely believed that legumes improve soil fertility because of the N_2 - fixing ability.

In nature, diazotrophy (N-fixation) is restricted to some prokaryotes and archaebacteria only. Diazotrophy is widely distributed in nature. One or the other organism exist for every ecological condition which can fix nitrogen either as free living or in association with higher or lower organisms. Under free living conditions, organisms are known which can fix nitrogen under anaerobic, microaerobic or aerobic conditions. Several diazotrophs fix nitrogen by forming symbiotic association with legumes and non-legumes and primitive members of plant kingdom. Symbiotic relationship could be perfect as in legume-rhizobia symbiosis or associative in which there is some interdependence between the partners but they can grow satisfactorily apart as in *Azospirillum* - cereal association. *Clostridium pasteurianum,* a strict anaerobe was the first organism shown to fix nitrogen. Most diazotrophs have 40°C as maximum temperature for nitrogen fixation, however, the archaebacteria *Methanococcus thermolithotrophicus* can fix at 65°C. Facultative anaerobes like *Klebsiella* can grow in both aerobic and anaerobic conditions, but for nitrogen fixation they need anaerobic environment as there is no mechanism of protection of nitrogenase against oxygen. Aerobic organisms like *Azotobacter* possess mechanisms to protect nitrogenase from oxygen.

Frankia forms nodules with higher plants like *Alnus* and *Casuarina.* Cyanobacteria can fix nitrogen in both aerobic and anaerobic conditions. The aerobic nitrogen fixation by cyanobacteria is either by special cells known as heterocysts which have anaerobic environment or by temporal separation of nitrogenase from oxygen during dark as observed in some unicellular cyanobacteria such as *Gloeothece.* Some cyanobacteria are symbiotic diazotrophs like *Anabaena azollae* and *A. cycadae.* With the advancement in knowledge of biological nitrogen fixation and advent of powerful techniques of molecular biology, the range of natural diazotrophs may get altered.

Symbiotic Nitrogen Fixers

In 1838, J.B. Boussingault conducted some experiments and found evidence that nitrogen enters legume plants by some unknown process, but not the non-legumes. This was questioned by many workers like J.Liebig, and others. In 1858, P.G. Lachmann studied the nodules in the roots of legumes and observed the bacteria. In 1866 M.Woronin further studied the nodule bacteria and found them to be rod-shaped and motile. In 1879, Frank showed that the root nodules are formed due to infection by the organism which he named *Rhizobium* (Plate 1). In 1884, W.O. Atwater conducted some studies with peas and came to the conclusion that the leguminous plants could fix free nitrogen of the air. In the same year, H. Hellriegal and H. Wilfarth brought out the connections between the capacity of legume plants to increase nitrogen content and the bacteria which are present in the nodules of these legumes, thus associating nitrogen fixation with legume nodules. However, the nature of the organism was understood only when M.W. Beijerinck in 1888 isolated the bacterium into pure culture and studied its physiology. The mechanism of root infection by the bacterium was studied by A.Prazmowski in 1889. Later workers like A.I. Virtanen *et al.* showed the differences in the amounts of nitrogen fixed in different plants and strain variations and cross inoculation groups of *Rhizobium* shown in table 1.

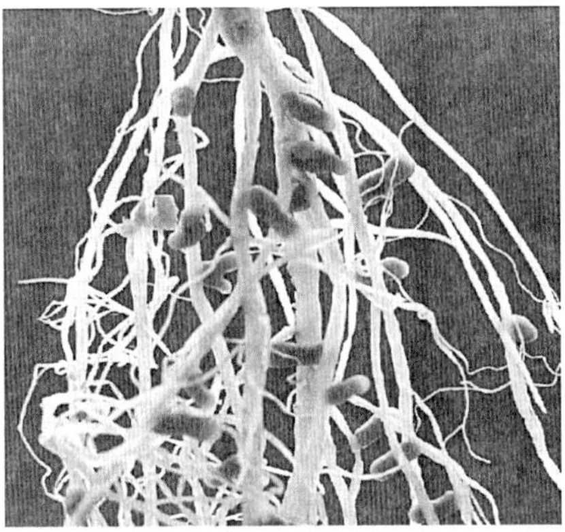

a

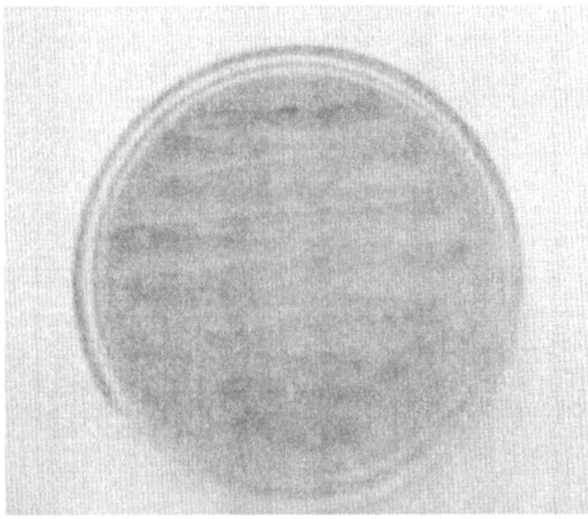

b

Plate 1 : a. Nodules of rhizobia in legume root, b. Rhizobia in plate

Table 1 : Cross inoculation groups of *Rhizobium*

Cross-inoculation group	*Rhizobium* species	Host it can nodulate
Alfalfa group	*R.meliloti*	*Medicago, Melilotus, Trigonella*
Clover group	*R.trifolii*	*Trifolium*
Pea group	*R.leguminosarum*	*Pisum, Vicia, Lathyrus, Lens*
Bean group	*R.phaseoli*	*Phaseolus*
Lupine group	*R.lupini*	*Lupinus, Ornithopus*
Soybean group	*R.japonicum*	*Glycine*
Cowpea	*Rhizobium* spp.	*Vigna, Crotolaria, Arachis, Cajanus, Dolichos, Pueraria*

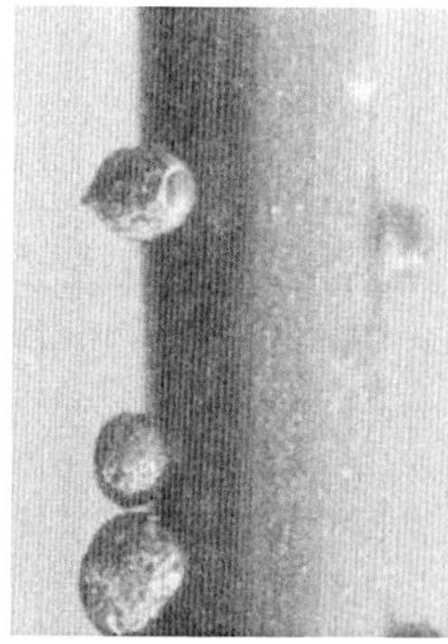

Plate 2 : Stem nodules on *Sesbania rostrata*

Certain rhizobia form stem nodules in legumes like *Sesbania rostrata* and *Aeschynomene indica* as observed by B.C Dreyfus *et al.* in 1985 (Plate 2). The capacity of stem nodules of *S. rostrata* to fix nitrogen is very high, exceeding 150 kg nitrogen fixed per hectare in 52 days. The bacterium has been named as *Azorhizobium caulinodans*. This system offers a unique potential as a green manure in rice cultivation. Rhizobia forming nodules in a non-legume *Parasponia* spp. has also been reported by M.J. Trinick in 1979. The symbiosis between the non-legume *Parasponia* and rhizobia is an extreme example of boundary jumping by rhizobia. Information on the efficiency and quantities of nitrogen fixed by different groups and species of symbiotic microorganisms is rather scattered. Also, wide variations in the quantities of nitrogen fixed by the same *Rhizobium-legume* combination under the influence of different agronomic and environmental factors have been reported. A compilation of the information on the subject is given in table 2. The quantities of nitrogen given are only approximate, cannot be taken as accurate, because the methods of estimation adopted by different workers may not be of the same value.

Table 2 : Approximate quantities of nitrogen fixed by different symbiotic systems

Host plant	Nitrogen fixed kg/hectare/yr
Leucaena leucocephala	500
Acacia mearnsii	200
Acacia pennatula	34
Inga jinicuil	35
Gliricidia sepium	13
Trifolium pratense	150-200
Glycine max	50-100
Pisum sativum	75-120
Vigna unguiculata	50-100
Medicago sativa	300-600
Pueraria phaseoloides	200-650
*Casuarina equisetifolia**	50-200
*Casuarina littoralis**	290
*Alnus glutinosa**	30-50
*Myrica gale**	8-10

* In respect of non-leguminous trees the nodulating organism is *Frankia* as against *Rhizobium* in legumes. *Frankia* is an actinomycete.

The nitrogen fixing bacteria in legumes, formerly known collectively as rhizobia, form nodules on roots (and exceptionally on stems) of some species of Leguminosae and on roots of *Parasponia* spp. (Ulmaceae). The term 'rhizobia' is related to Rhizobiaceae Conn 1938, a family-level taxon of bacteria created expressly to accommodate organisms able to nodulate Leguminosae. A significant recent extension of knowledge regarding nitrogen fixing Leguminosae nodulating bacteria was the discovery that bacteria belonging to the β-proteobacteria (genera *Burkholderia* and *Ralstonia/ Cupriavidus*) and to other families in the α-proteobacteria (e.g. *Methylobacterium,* Methylobacteriaceae and *Devosia,* Hyphomicrobiaceae) were also able to nodulate Leguminosae (Table 3). For this reason the name 'rhizobia' is no longer suitable to describe nodulating bacteria, although its use continues in some literature.

Leguminosae comprise about 20,000 species distributed in the subfamilies Caesalpinoideae (2,250 spp., mainly woody tropical plants), Mimosoideae (3,270 spp., mainly woody tropical, sub tropical and temperate) and Papilionoideae (13,800 spp., largely herbaceous). The extent of the symbiosis with nitrogen fixing legume nodulating bacteria (NFLNB) among Leguminosae is still the subject of active research. Until 1989, only 57 per cent of genera and 20 per cent of species have been examined for nodulation, and the percentage of species able to nodulate were 23, 90 and 97 among Caesalpinoideae, Mimosoideae and Papilionoideae respectively. Although extensive searches for new nodulating genera and species have been made, especially in Brazil, it is estimated that today only 25 per cent of the extant species have been examined. It is thus possible that many new symbioses will be identified in natural ecosystems (e.g. tropical forests). Traditionally, the taxonomy of NFLNB was based on strains isolated from temperate crop plants. However, now that isolates from other species and regions are available, and now techniques of molecular biology have been developed, the taxonomy has been revolutionized, with 51 species and ten genera added to the original four species and two genera listed in 1984 by Jordan (Table. 3) and further revisions are expected.

Table 3: Phylum/ order, family, genera and species of legume nodulating bacteria.

Phylum/family/genera/species
α-proteobacteria- Order Rhizobiales
Rhizobiaceae
Rhizobium
R. leguminosarum, R. galegae, R. tropici, R. etli, R.giardinii, R. gallicum, R. sullae, R. hainanense, R. mongolense, R. huautlense, R. indigoferae, R. loessense, R. daejeonense, R. yanglingense
Sinorhizobium
S. meliloti, S. fredii, S. xinjiangense, S. saheli, S. teranga, S. medicae, S. arboris, S. kostiense, S. adhaerens, S. morelense, S. kummerowiae, S. americanum
Allorhizobium
A. undicola
Bradyrhizobiaceae
Bradyrhizobium
B. japonicum, B. elkanii, B. liaoningense, B. yuanmingense, B. canariense
Blastobacter *(Bradyrhizobium)*
B. denitrificans
Xanthobacteraceae
Azorhizobium
A. caulinodans, A. doebereinerae.
Phyllobacteriaceae
Mesorhizobium
M. loti, M. huakuii, M. ciceri, M. tianshanense, M. plurifarium, M. amorphae, M. chacoense, M. temperatum, M. spetentrionale, M.mediterraneum
Phyllobacterium
Phyllobacterium trifolii
Methylobacteriaceae
Methylobacterium
M. nodulans
Brucellaceae
Ochrobactrum
Ochrobactrum lupini
β-Proteobacteria Order **Burkholderiales**
Burkholderiaceae
Burkholderia
B. tuberum, B. phymatum, B. caribensis, B. cepacia genomovar VI, *B. mimosarum*
Ralstonia (Cupriavidus)
R. (Cupriavidus) taiwanensis
Hyphomicrobiaceae
Devosia
D. neptunea

Evaluation of NFLNB diversity in soil

Studies of NFLNB populations and NFLNB diversity depend on successful isolation from root (or occasionally stem) nodules of legume host plants. Nodules may be sampled from field grown plants (*in situ* collections), although this approach may be difficult with woody or perennial hosts. In order to isolate and enumerate NFLNB from diverse microbial population, such as that which occurs in the soil, a method is required that clearly separates NFLNB from other bacterial species and allows easy and systematic sampling of nodules from nodulated plants. The plant infection technique makes use of the nodulation process itself to estimate NFLNB populations in soil. Further culture and characterization of the NFLNB from the nodules thus formed can provide information on the taxonomic composition of NFLNB populations and degrees of specificity between particular strains and candidate hosts. Although some host plants are considered highly promiscuous (*i.e* show low NFLNB specificity), no one promiscuous host can be nodulated by all existing NFLNB species / strains, and conversely, there is no existing NFLNB strain sufficiently promiscuous to nodulate all legume species. Thus, to evaluate NFLNB diversity in soil it is desirable to make use of variety of candidate host plant species, and the more that are employed, the greater will be the variety of NFLNB strains isolated and identified. The bioassay for NFLNB can make use of promiscuous hosts grown in soil sampled from the field or inoculated with soil suspensions and the species found should then be compared with species nodulating naturally at the site. Although some of the later associations may be relatively specific, the comparison of NFLNB isolated from naturally formed nodules with those sampled *via* the bioassay provides a useful check on the accuracy of the laboratory procedure. For instance, *Macroptilium atropurpureum* is one of the widely accepted promiscuous hosts (Vincent, 1970), but in most cases it is reported as being predominantly nodulated by *Bradyrhizobium* species, although it can also be nodulated by fast growers such as *Rhizobium* spp. and *Burkholderia* spp., among others (Moreira, 2008). Preliminary results from Kenya, indicated that *M. atropurpureum* was predominantly nodulated by slow growing *Bradyrhizobium* spp. Similarly, Lewin *et al.* (1987) demonstrated that *Vigna unguiculata,* usually considered a *Bradyrhizobium* host, in fact has very low specificity and can be nodulated by fast growing *Rhizobium* spp. Results from Brazil showed that cowpea traps more species than *M. atropurpureum* and beans (*Phaseolus vulgaris*), but only siratro trapped

NFLNB from all the sampling points. Preliminary experiments are therefore recommended for each soil environment before the final choice of host plants is made for the bioassay; where possible a native legume species should be included as trap host.

The establishment of legume-*Rhizobium* symbiosis is a complex process involving interplay of physiological and biochemical processes of both bacterium and the host plant. The symbiosis occurs in a host specific way involving several steps of signal exchange between plant and bacterium, which leads to formation of nodules. In the rhizosphere of legumes, rhizobia move towards legume roots due to the presence of chemo-attractants in root exudates. The nature of the substances present in root exudates showing positive chemotaxis, depends upon the legume host. Various rhizobia are attracted by amino acids, dicarboxylic acids, flavonoids and betaines. After the chemotaxis, rhizobia attach to the plant root surface. The most favoured sites of attachment are the young growing root hairs, but attachment can take place on root hairs at other sites also. Initially, rhizobial attachment takes place loosely as single cell and later several cells firmly attach to form a cap structure. The attachment may involve several macromolecules including rhizobial extracellular polysaccharides, lectins of legume roots and calcium dependent adhesin called rhicadhesin. The firm attachment of rhizobia in the cap structure may require cellulose fibrils or proteinaceous bacterial fimbriae with plant lectins. When rhizobia colonize legume roots, deformation of root structure takes place. Deformations include thick and short root formation, root branching and curling of root hairs. These changes in root morphology are also brought out by application of culture supernatant (Nod factor) of rhizobia induced with root exudates.

There are three different methods by which rhizobia infect and enter the legume roots. These include infection thread formation, crack entry and penetration at the junction of epidermal cells. The mechanism of infection and the mode of entry is host specific and is independent of infecting rhizobial strain. In the formation of infection thread, *Rhizobium* attached to the roots, causes curling of root tips and gets entrapped in the curls. This results in a localized hydrolysis of plant cell wall which causes invagination of plasma membrane. At the site of invaginated plasma membrane, a tubular structure is formed due to the deposition of new cell wall material. The tubular structure is known as infection thread through which bacteria enter the roots. Concomitantly with the formation of infection thread, particular cortical cells divide mitotically forming the nodule primordia. In temperate legumes, having indeterminate cylindrical nodules, the nodule primordium is derived from inner cortical cells, while in tropical legumes having determinate nodules, it is derived from outer cortical cells.

Nodule formation by rhizobial entry through cracks in the epidermis is observed in peanut and stem and roots of *Sesbania rostrata.* In peanut, rhizobia induce cell division in the cortex of young lateral roots, causing separation of cortical and epidermal cells that enables entry and intracellular spread of rhizobia. No infection thread is formed and probably by cell wall digestion, outer root cells collapse and due to continuous cell division, central tissue forms nodules. In *Sesbania,* direct intra cellular entry of rhizobia is followed by multiplication of rhizobia forming narrow, branched intracellular infection thread. The infection thread is extended to meristematic zone and nodule is formed. In *Parasponia andersonii,* the only non-legume host forming nodules with *Rhizobium,* the entry of rhizobia is through wounds caused by rupture of epidermis at the base of root hairs. The epidermis ruptures due to the formation of callus like swollen structure induced by the cell proliferation caused by rhizobia.

After the tissue invasion, bacteria are released into the plant cell cytoplasm enveloped in a peribacteroid membrane (PBM) and expression of most bacterial genes stops. The bacteria differentiate into bacteroids and each bacteroid is surrounded by PMB derived from plant cytoplasmic membrane. Bacteroids divide within the PMB and finally form a structure known as symbiosome which may contain up to 20 bacteroids. In the symbiosomes, nitrogenase enzyme complex is synthesized which catalyzes the reduction of dinitrogen. The space surrounding bacteroids in a symbiosome is known as peribacteroid space filled with the pigment leghaemoglobin. In the nodule, majority of the cells are infected with *Rhizobium,* but a small fraction of uninfected cells is also present. Both the infected and uninfected cells of nodules play important role in nodule functioning. Most of the functions related to dinitrogen reduction take place in the infected cells, while function related to ammonia assimilation and production of carbon metabolites are performed by the uninfected cells. The amount of leghaemoglobin and bacteroid production has direct relationship with the amount of N_2 fixed by legumes.

During nodule development and maintenance, a number of plant and bacterial 'genes' are expressed sequentially in a well coordinated manner. Nodulins are nodule specific protein encoded by plant genes and produced during certain stages of nodule development. Based on the possible functions, these may be divided into three categories: (i) proteins responsive for maintenance of nodule structure. (ii) enzymes induced in specific nitrogen assimilates and carbon metabolism of nodules. (iii) proteins that support bacteroids function and thus facilitate nitrogen fixation. Based on the structural similarities of nodulins, two groups are recognized - one is C-nodulins or common nodulins as the name indicates- protein common to all nodules - and the other is S-nodulins or species- specific nodulins.

Factors affecting legume- *Rhizobium* symbiosis

Ecological factors

Soil temperature at the sowing depth affects the efficiency of rhizobia. In tropical and subtropical countries like India, the temperature during summer month is as high as 45 - 50°C. High temperature affects root exudation, growth and survival of rhizobia, root- hair formation in plants and infection process.

Moisture or water stress limits not only the survival of rhizobia but also their symbiotic association in the legumes. It has been reported that water stress (-2 to -4 bars) resulted in decreased growth of rhizobial strains.

The process of symbiotic nitrogen fixation is adversely affected by combined N, especially NO_3. The inhibitory effects of nitrate on infection, nodule development and N_2 fixation are well known. The exact mechanism of inhibition is not very clear. The theories proposed ascribe the inhibition to a diminished supply of photosynthate available to nodules following its use for assimilation of NO_3 in the shoots and roots; inhibition of root hair curling due to catalytic destruction of IAA; inhibition of MoFe protein of nitrogenase due to binding of the NO_3, decreased production of lectin and the formation of attachment sites for these lectins on the root hairs (Dazzo *et al.*, 1978).

The high concentration of salts in the soil has a detrimental effect on host and rhizobia and their symbiosis. Legumes are generally more sensitive to osmotic stress than their microsymbionts. Rhizobial strains from different groups have been reported to tolerate 0.5 - 5% NaCl concentrations, both in culture and in the soil. Generally, fast growing rhizobia are reported to tolerate NaCl levels as high as 0.34 M, which the slow growing soyabean bradyrhizobia failed to cope with.

Soil acidity constrains symbiotic nitrogen fixation in both tropical and temperate soils limiting *Rhizobium* survival and persistence in soils and reducing nodulation. It has been reported by several workers that the presence of organic matter in soil has a favourable influence on the number of rhizobia, nodulation and nitrogen fixation.

Nutritional factors affecting nodulation include P, N, K, Mo, Zn, Fe, Mg, S, Co, Ca, Cd, Mn and Cu. Phosphorus deficiency is a factor that commonly restricts the realization of the potential of N_2 fixation by legumes. Among secondary nutrients, calcium and magnesium play an important role in legume-*Rhizobium* symbiosis. Calcium is involved in cell-wall synthesis of *Rhizobium*, and if grown in calcium deficient medium, it showed decreased attachment ability. Among micronutrients there is evidence that the requirement of boron for nodule development is similar to that for growth of the host. Likewise for nitrogenase activity, the need for Mo and Co far exceeds other plant requirements. Zn^{++} is required for the production of indole acetic acid, which is important for nodulation in legumes.

Biotic factors

The presence of numerous parasites and predators poses problems in the establishment of rhizobial biofertilizer. The presence of antagonistic microflora has been postulated to affect the out come of competition between the strains of rhizobia. *Erwinia herbicola,* an epiphytic bacterium is commonly associated with seeds of *Medicago sativa.* It has been shown to inhibit nodulation by *R.meliloti* by producing toxin or by blocking rhizobial attachment sites on *M.sativa* root hair. Rhizobial phages and *Bdellovibrio* in soil can also infect and lyse rhizobia in soil.

Frankia

The actinomycete *Frankia* is another nodulating nitrogen fixing organism forming nodules in non-legumes (Plate 3). *Frankia* induce N_2 - fixing root nodules, actinorhiza, on more than 280 species of woody plants from 8 different families (Schwintzer and Tjepkema, 1990). In temperate region, alders (*Alnus* spp.) have great potential in various types of forestry and documented to have considerable N_2 fixation even at young age. *Casuarina,* another non-legume is the most important actinorhizal genus of the tropics and subtropics. Their ability to form symbiotic associations with *Frankia* enables them to thrive in N- deficient soils and makes them desirable for agroforestry. Some occur as regular components of climax vegetation while others occur as pioneer species after disturbance or deposition of fresh soil deposits such as volcanic ash, glacial till or sand dunes.

Plate 3 : *Frankia* nodules in *Casuarina* root

Taxonomy and systematics

The criteria used in classifying an actinomycete as a member of the genus *Frankia* include the following: cell morphology, sporangia and vesicle formation (in submerged liquid cultures); chemistry (cell wall of type III, phospholipid of type PI); serology, DNA homology and demonstration of infectivity and / or effectivity for a host. Thus, the new taxonomy in Bergey's Manual of Systemic Bacteriology Vol. IV tries to reflect the biochemical and cell physiological properties of the bacteria as well as specific plant reactions and symbiotic specificity.

Physiology

The biology and ecology of *Frankia* is not well understood. This is partly due to difficulties encountered in the past in the isolation of this organism in pure cultures. Reasons for difficulties in isolating the symbiotic *Frankia* are:

(i) The very slow growth of these bacteria in relation to other actinomycetes and eubacteria.

(ii) The preponderance of rapidly growing contaminating eubacteria and actinomycetes associated with soil-borne structure.

(iii) The unknown media preferences of *Frankia* making the transition from a symbiotic to a heterotrophic or saprophytic metabolism.

(iv) The inhibiting nature of phenolic compounds released from host plant cells during isolation procedures.

Since the first success in reproducible isolation of the endophyte from actinorhizae in 1978 (Callaham *et al.*, 1978), a number of *Frankia* isolates have been obtained. Most of the *Frankia* strains isolated from *Alnus* and *Myrica* species are morphologically almost indistinguishable, when grown under standard culture conditions. They form similar vegetative hyphae while sporangia are produced in variable amounts from strain to strain. DNA hybridization studies showed a high degree of homology between *Alnus* and *Comptonia* nodules. However, isolates from *Elaeagnus* and *Casuarina* species differ from each other and also from the *Alnus* types. Based on their physiological reactions, all these isolates have been categorized into three groups *viz.*, A, B and C. Groups A and B have been studied more as compared to C. Table 4 lists the differences between A and B groups.

Table 4: Differences between type A and B groups of *Frankia*

Character	Type A	Type B
Cellular pigments	+	-
Vesicles	variable (on complex media)	+ (on N-free media)
Synergism (Tween + carbohydrate)	-	+
Depressed growth (Tween + carbohydrate)	+	-
Carbohydrate uptake at 0.5%	+	-
Maintenance on slants	+	-
Infective on original host	-	+ (usually)
Serological group	II or other	I
Whole cell sugar	Various: fucose, mannose xylose, glucose, galactose	Xylose
Plasmid	Labile	Stable

The optimal growth temperature for *Frankia* is 30°C. No growth is observed at 14°C or above 36°C. Differences in carbon metabolism are known for *Actinomyces* and similar differences are also found among the members of *Frankia*. Some strains like those from *Comptonia peregrina* grow in defined medium using succinate and glutamine as carbon source, where as strains from *Alnus incana* grow using acetate and propionate and are incapable of using succinate, fumarate, malate or glucose. However, a general simple medium with all macronutrients, trace elements and propionic acid as carbon source can be used for all *Frankia* isolates.

In contrast to the specificity exhibited in *Rhizobium* - legume symbiosis, the host spectrum of *Frankia* is very wide. These non-legume hosts belong to 20 different genera from 8 families and 7 different orders. About 160 species of these members are known to develop actinorhizal nodules. The most important genera are *Allocasuarina, Alnus, Elaeagnus, Myrica, Gymnostoma, Casuarina* and *Coriaria* (table 5).

Table 5 : Actinorhizal families, genera and species

Family	Genus	Number of species
Betulaceae	*Alnus*	47
Casuarinaceae	*Allocasuarina*	54
	Casuarina	16
	Ceuthostoma	2
	Gymnostoma	18
Coriariaceae	*Coriaria*	16
Datiscaceae	*Datisca*	2
Elaeagnaceae	*Elaeagnus*	38
	Hippophae	2
	Shepherdia	2
Myricaceae	*Comptonia*	1
	Myrica	28
Rhamnaceae	*Ceanothus*	31
	Colletia	4
	Discaria	5
	Kentrothamus	1
	Retanilla	2
	Talguenea	1
	Trevoa	2
Rosaceae	*Cercocarpus*	-
	Charmaebatia	1
	Cowania	1
	Dryas	1
	Purshia	2

Inoculation groups

Three major host infectivity groups using fifty different strains of *Frankia* have been defined. These are: (i) strains which nodulate *Alnus* and *Comptonia*. (ii) strains which nodulate *Casuarina* and *Myrica*. (iii) strains which nodulate Elaeagnaceae and *Myrica*.

Structure and function of actinorhiza

Actinorhizae develop only when soil around the roots is well aerated. Dryness and high soil temperature inhibit infection. Under limiting conditions the effect of symbiotic N_2 fixation on host growth is markedly observed. Most of the research on nodule infection cytology has been carried out in *Alnus*. As in legumes, the root hair is the site of primary infection in *Alnus*.

Structure of actinorhizal nodule

There are two types of structural organization in actinorhizal root nodules - the *Alnus* type and the *Casuarina* type. *Alnus* type of nodules have many lenticels on nodules providing ventilation while *Casuarina* type do not have ventilatory organs. Instead, those lateral roots that escape infection by *Frankia* develop into nodule roots serving as antennae for oxygen diffusion.

Use of nitrogen-fixing trees for improved forest productivity has enormous potential especially where long practices have been designed so as to maintain soil organic matter and water holding capacity. Present use of actinorhizal plants includes re-colonization of disturbed land, in shelter belts and forestry interplanting. Values between 50 and 200 kg N fixed/ha/year have been reported for *Alnus glutinosa* and *A.rubra*. Economically actinorhizal plants such as *Casuarina* in tropical countries and *Alnus* or *Elaeagnus* in temperate countries are interesting for reforestation and reclamation of mine spoils, dune stabilization and timber production.

Associative Nitrogen Fixers

Bacteria closely associated with plants inhabiting rhizosphere, rhizoplane and inner cells of plants and capable of fixing atmospheric nitrogen are referred to as associative nitrogen fixers.

The first associative diazotroph was reported by Beijerinck in 1925 under the name *Spirillum lipoferum*. However, it was about half a century later, after the discovery of the highly specific *Azotobacter paspali-Paspalum notatum* association and the rediscovery of *Spirillum lipoferum* (now called *Azospirillum*) by the group of Dobereiner (Dobereiner *et al.*, 1997), that scientists became increasingly interested in diazotrophic bacteria associated with graminaceous plants.Because the benefit of nitrogen fixation from nodulated legumes to agriculture was already established at that time, it was expected that the associative diazotrophs would favour non-leguminous plants in the same way. Several genera of bacteria have now been reported to contain diazotrophs, which may be loosely or more initimately (endophytes) associated with plant, including *Gluconacetobacter, Azoarcus, Azospirillum, Herbaspirillum,* etc. An extensive phylogenetic classification of nitrogen fixing organisms was made by Young in 1994. While the capability of these organisms to fix nitrogen *in vitro* can be demonstrated easily, efforts to quantify nitrogen fixation in natural associations with plants have produced widely varying results. In recent years a new orientation to the study of associative nitrogen fixers has come.

Due to its microaerophilic nature, *Azospirillum* can be isolated on a semi-solid malate containing medium by enrichment procedures (Plate 4a). The development of white, dense and undulating fine pellicles on a semi-solid malate medium is a characteristic feature of *Azospirillum.* During enrichment the dominant organisms on a sodium malate medium are characteristically curve rods of various sizes with prominant refractive fat droplets. This organism is Gram-negative and contains poly-β hydroxy butyrate granules. Microscopic examination reveals polymorphism and spirillar movement. For fixation of molecular nitrogen, the bacterium needs microaerophilic surroundings (low O_2 conditions) but can grow profusely on ammonium containing medium without fixing nitrogen. The ability of the organism to fix nitrogen has been verified by the acetylene reduction test and uptake of $^{15}N_2$ gas. In terms of mg N_2 fixed/g of substrate added, the range of fixation, as measured by Kjeldahl method, varies from 12 to 36 among different isolates of *A. brasilense.* The organism is also known to produce growth substances such as IAA, kinetins and gibberlins.

Positive effects of *Azospirilllum* inoculation on growth and yield of several crop plants, mainly cereals have been reported and reviewed from time to time. Yield increase accompanied by increased nitrogen concentration following bacterial inoculation might be due to enhanced nitrogen fixation and/or production of plant growth promoting substances.

Herbaspirillum

Herbaspirillum sp. is an endophytic diazotropic bacteria associated with important agriculture crops. It is isolated from roots of maize, rice, soyabean, forage grasses and recently from leaves, stems and roots of sugar cane. *Herbaspirillum* cells are Gram-negative, aerobic, curved, motile rods that secrete enzymes as pectinase and cellulase. This endophytic diazotroph invades roots by adsorption and proliferation and secondary roots and root surface damage. It colonizes intercellular spaces in young tertiary leaves and coleoptiles and prefers apoplast of shoot tissues. Increased biomass/yield up to 31% has been reported because of *Herbaspirillum* inoculation.

Gluconacetobacter

A diazotrophic bacterium, *Acetobacter diazotrophicus,* now renamed as *Gluconacetobacter diazotrophicus,* mainly associated with sugar-rich plants like sugar cane, sweet potato, Cameroon grass, sweet sorghum, has been reported. The organism has also been isolated from coffee and finger millet. It colonizes roots, stems and leaves of sugarcane in numbers up to 10^6 cells/g fresh weight and that of sweet potato up to 10^5/g.

G. diazotrophicus is a Gram-negative, acid tolerant, obligate aerobe, straight rod with curved ends, occurring in pairs/chains, non-endospore former, opportunistic endophyte isolated from roots and stems of sugar cane and not from root free rhizospheric soils. The bacterium is highly saccharophilic as it grows at 10% sucrose concentration and at low pH (3.0) (Plate 4b). It secretes extracellular levansucrase, is antagonistic to *Colletotrichum falcatum,* and is able to solubilize phosphorus in addition to fixing nitrogen. This bacterium persists from one generation to the next of the vegetatively propagated plant. Nitrogen fixed by *G. diazotrophicus* is excreted as ammonia into the medium. However, no correlation was observed in the number of bacteria and the capacity of sugarcane varieties to obtain N from BNF. Strains of *Gluconacetobacter* have been shown to produce considerable amount of IAA. Several varieties of sugarcane in Brazil are capable of obtaining over 60% of their nitrogen (=150 kg N/ha/year) from BNF with the association of *G. diazotrophicus*. Field trials in India have shown that nearly 50% of fertilizer N can be reduced when sugarcane crop is inoculated with *G. diazotrophicus.*

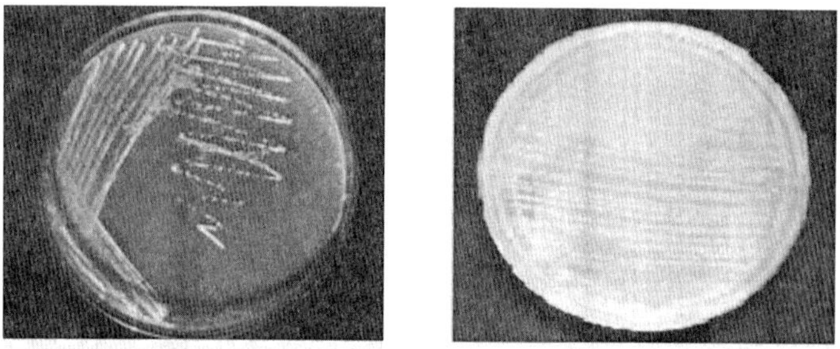

a b

Plate 4 : Plates showing growth of a. *Azospirillum brasilense*, b. *Gluconacetobacter diazotrophicus*

Free -living Nitrogen Fixers

Free-living nitrogen fixers grow in soil or in the rhizosphere and they do not enter the plant roots. Though there are a number of free living nitrogen fixing bacteria reported in literature the most important group is the heterotrophic bacteria belonging to the family Azotobacteriaceae that include three genera *viz.*, *Azotobacter*, *Beijerinkia* and *Derxia*, and another important group being the free living cyanobacteria.

Azotobacter

As compared to other bacteria, *Azotobacter* is large sized (ranges from 2-10 x 1-2.5 μm) Gram-negative, pleomorphic (different sizes and shapes) bacteria, producing copious amounts of polysaccharides and motile due to peritrichous flagella. Some species of this genus produce pigments, yellow-green, red-violet, or brownish-black while extracellular slime production is a common feature of genus, that helps in sequestration of water in the immediate vicinity and retain it (Plate 5). Old cultures of *Azotobacter* exhibit presence of encapsulated forms 'cysts' that possess enhanced resistance to heat, desiccation and adverse conditions while germinate on advent of favourable conditions to form vegetative cells. *Azotobacter* sp. are sensitive to acidic pH, high salts, and temperature above 35°C. This genus includes several species, which could be distinguished on basis of cell size, pigmentation and extra cellular slime production. Cyst formation in *Azotobacter* is a special character but this characteristic debars the bacterial nitrogen fixing potential. Moreover these cysts do not behave as spore (for being not totally dormant). *Azotobacter* uses carbon for its metabolism from simple or compound substances of carbonaceous nature and requires calcium, while it produces thiamine, riboflavin, nicotin, indole acetic acid and gibberllin. *Azotobacter* strains possess varied chemical and biological characteristics equipping some strains to fix atmospheric nitrogen with higher efficiency than the others.

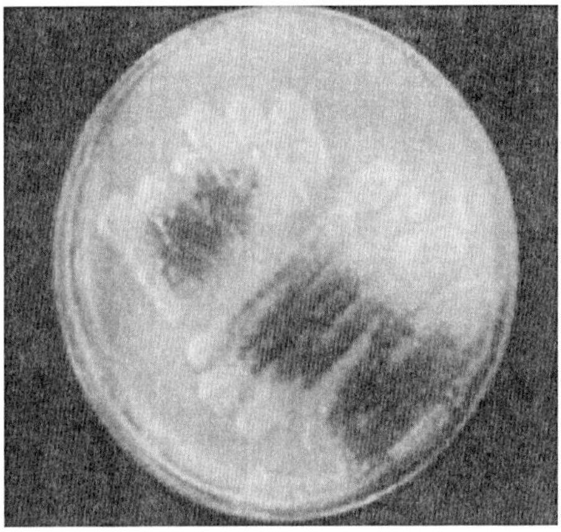

Plate 5 : Plate showing growth of *Azotobacter chroococcum*

The diazotrophic nature of *Azotobacter* is due to its ability to fix atmospheric nitrogen by the conversion of elemental (N_2) form to ammonia (NH_3) accomplished by virtue of nitrogenase enzyme complex. The nitrogenase enzyme diversity and extremely rapid metabolic rate (the highest of any known living organism) allows this bacterium to fix nitrogen under aerobic conditions while other nitrogen-fixing bacteria possess only a single species of nitrogenase, which requires anaerobic to microaerophilic conditions in order to function. This enhanced versatility of *Azotobacter* (regarding nitrogenase enzyme complex) makes the microbe attractive for agricultural purposes. They fix about 15-30kg N/ha /year depending upon the availability of organic matter in soil.

Presently the genus *Azotobacter* has six species *viz. A. chroococcum, A. vinelandii, A. beijerinckii, A. nigricans, A. armemacus* and *A. paspali.* Except the last species, which is an exclusively rhizospheric bacterium, the others are largely soil inhabitants. *Azotobacter* being aerobic, needs continuous supply of oxygen for its function. The optimum temperature for their growth is between 25°C and 30°C and the optimum pH for the growth is near to slightly above neutrality (7.0 to 7.6). Salinity and nitrogenous fertilizers inhibit their growth and nitrogen fixation.

Beneficial effect of *Azotobacter* inoculation has been reported by various workers on range of crops including cereals, pulses, vegetables, and oilseed. Of the different species of *Azotobacter, A. chroococcum* is being used as inoculum for seed bacterization of agricultural crops. The multiple action of *Azotobacter* contributes to better germination percentage of seeds, increased shoot and root length, improved N nutrition, reduction in disease incidence, increased grain yield and improved post-harvest seed quality in terms of germination.

Beijerinckia

This free living, aerobic, chemoheterotrophic nitrogen fixer exhibits characteristic features *i.e.* formation is copious amount of slime, rod-shaped cells with round ends having polar lipoid bodies, nitrogen fixation at very acidic pH values (3.0-4.0) which differentiate this genus from *Azotobacter.*

Derxia

This genus contains a single species *D. gummosa* characterized by presence of pleomorphic, motile (by short polar flagellum), Gram-negative bacilli occurring singly or in short chains, capable of nitrogen fixation under aerobic or microaerobic conditions. Its efficiency of nitrogen fixation varies between 9 and 25 mg N/g of glucose consumed, but in most strains it is distinctly lower than in *Azotobacter* or *Beijerinckia.*

Cyanobacteria

Cyanobacteria earlier referred to as blue green algae are unicellular, colonial or filamentous forms occurring mostly in aerobic and some times in microaerophilic environments. Cyanobacteria belonging to the orders Nostocales and Stigonematales are the most common nitrogen fixers and include the genera *Anabaena, Anabaenopsis, Aulosira, Chlorogloea, Cylindrospermum, Nostoc, Calothrix, Scytonema, Tolypothrix, Fischerella, Haplosiphon, Mastigocladus, Stigonema* and *Westiellopsis* (Fig. 14). There are two general types *viz,* the coccoid and filamentous types. The coccoid species occur as single cells, colonies or masses of various shapes (Plate 6). The colonies in mat remain enclosed in a gelatinous sheath that varies in consistency and thickness. The filamentous types generally possess the nitrogen fixing attribute which is characteristic of heterocystous cyanobacteria but exceptions to this have been reported where non-heterocystous types exhibited nitrogen fixation. Filamentous cyanobacterial cells are characteristically arranged in rows to form trichomes that are then enveloped in a common enclosing sheath to form filaments.

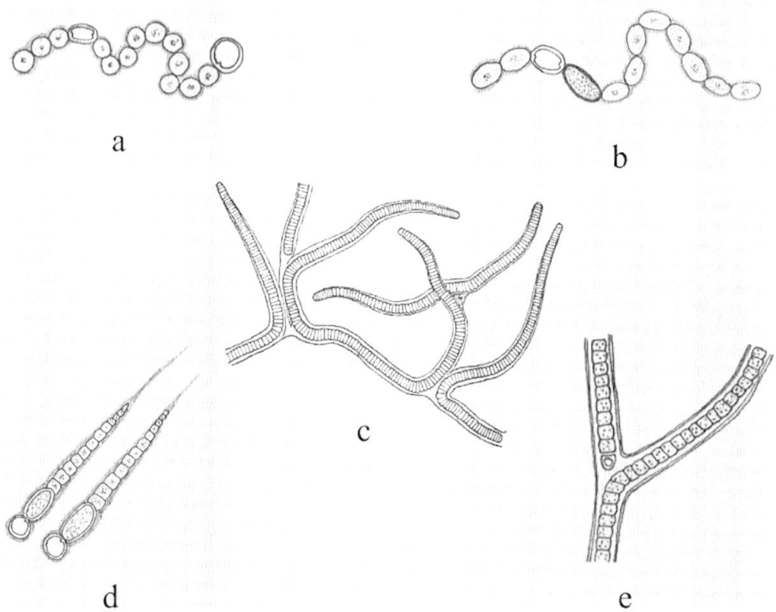

Fig. 14 : Some common cyanobacteria occurring in soil
a. *Nostoc,* b. *Anabaena*, c. *Plectonema* , d. *Rivularia*, e. *Tolypothrix*

Some filamentous species exhibit cell differentiation and form heterocysts, characterized as large cells heterotrophic in nature, possessing extraordinarily thick cell wall connected by cytoplasmic bridges to neighbouring photosynthetic cells, lack an oxygenic photosystem and lack biliprotein pigments and carboxysomes thus appearing empty amidst normal pigmented vegetative cells. Heterocysts provide congenial environment for effective functioning of nitrogenase and are considered as the sites of nitrogen fixation. The number of heterocysts could be taken as a rough parameter to indicate the nitrogen fixing capacity of cyanobacteria as amount of nitrogen fixed depends upon physiological and environmental factors. Apart from forming heterocysts some heterocystous cyanobacterial species form akinete cell type that germinates on advent of favourable conditions to form new vegetative cells for multiplication and formation of a new colony.

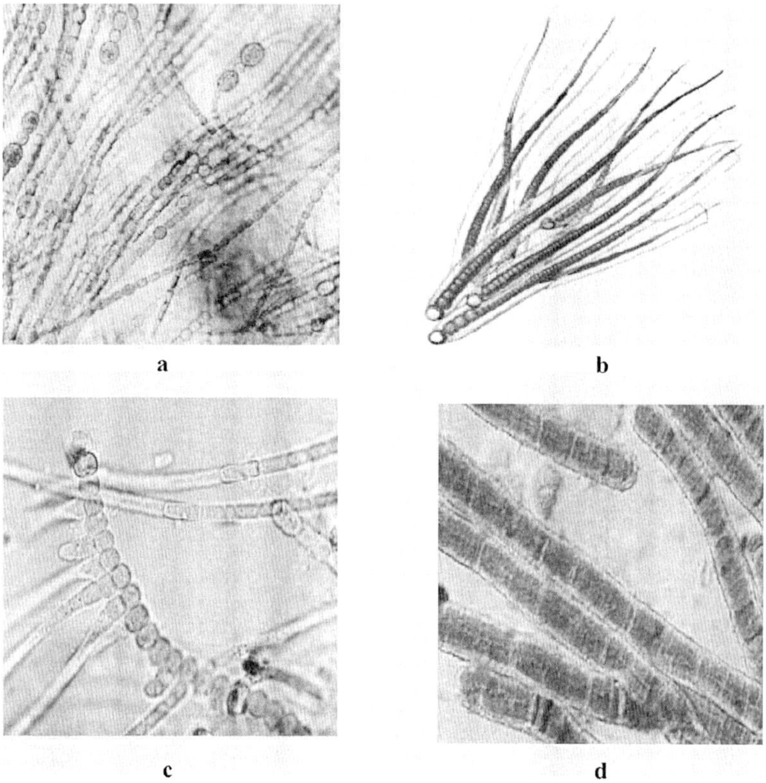

Plate 6 : a. *Nostoc* b. *Rivularia* c. *Fischerella* d. *Scytonema*

Most cyanobacteria develop best in neutral or weakly alkaline media. Normally below pH 6.5 their growth is adversely affected. Cyanobacteria have a very distinctive response to oxygen and this warrants their being categorized as aerobes. The optimum temperature for blue green algae is close to 28°C - 30°C. Sunlight is another very important factor for their development. Most of these microorganisms seem to be obligate phototrophs. The correlation between photosynthesis and nitrogen fixation has been observed. The vigour of nitrogen assimilation is also influenced by the concentration of CO_2 in the medium.

Cyanobacteria play on important role in the accumulation of nitrogen in rice fields where conditions are optimum during a considerable part of the growing season. Most investigators have reported that in these conditions the annual nitrogen increment due to the activity of cyanobacteria varies between 15 and 50 kg/ha which may go up to 80 kg/ha sometimes.

Characteristics of Nitrogen Fixation

Three conditions must be met to prove that bacterial N_2 fixation occurs. First, there must be a large population of these organisms that is consistent with the observed N_2 fixation rate. Second, there must be rapid cell formation, which indicates that N_2 fixation is linked to growth. Third, the N_2 must be atmospheric, not inorganic or organic; this is one of the most difficult conditions to meet.

Nitrogen fixation is energy intensive. Microorganisms must have a ready supply of electrons for significant N_2 fixation to occur. Electrons are necessary because nitrogen fixation is a reductive process. Nitrogen gas (valence state of 0) gains electrons to become NH_3 (valence state-3). Reduced C is used by heterotrophs to supply the electrons while light energy is used by phototrophs to oxidize either water or a reduced sulphur compound to provide the necessary electrons in anaerobic conditions.

Nitrogen fixation requires an enzyme complex called nitrogenase, and nitrogenase must operate in a microenvironment protected from O_2. Nitrogen fixers must find or produce an O_2 - free or O_2 - limited environment. This is a dilemma for aerobic nitrogen fixers such as *Azotobacter* because O_2 respiration is required to produce electrons for fixation and for respiration. There are several mechanisms that are used to limit O_2. *Azotobacter* can limit O_2 diffusion by producing copious amounts of slime. It can limit O_2 by developing high respiration rates and consuming O_2 as rapidly as O_2 diffuses in. *Azotobacter* has the highest respiration rate of any microorganism known, 500 to 1,000 cm^3 O_2 consumed per g cells. *Azotobacter* also makes reversible changes in the arrangement of nitrogenase to convert it from an O_2 - sensitive functional form to an O_2 - tolerant, but nonfunctional, form.

Anaerobes such as *Clostridium* are protected from O_2 because they exist in an anaerobic environment. However, energy generation is a problem. Consequently, the amount of N_2 fixed globally by anaerobes is small but significant where there are no other external sources of nitrogen.

Phototrophs, such as cyanobacteria, make O_2 much as plants do. The ability to fix N is most common in filamentous heterocyst-forming cyanobacteria. Nitrogen fixation occurs in the heterocysts, which are thick- walled, multilayered, vegetative cells that lack photosystem II. Consequently, no O_2 is produced during photosynthesis in these cells.

Photosystem II appears only in non-heterocystous, vegetative cells. Cyanobacteria that do not form heterocysts can fix N_2 only when they are not photosynthesizing.

Nitrogen fixation does not occur when NH_4^+ or NO_3^- or organic N are readily available. A number of hypotheses have been offered to explain this. The presence of available N may direct electron flow away from nitrogenase - no electrons; no reduction. Another theory invoked for symbiotic systems is that NO_3^- reduction in nodules causes NO_2^- binding of compounds that protect nitrogenase from O_2.

Physiology of Nitrogen Fixation

Biological N_2 fixation is catalyzed by an enzyme association called nitrogenase. This is strictly a prokaryotic enzyme complex. Nitrogenase is composed of two soluble proteins: the Fe protein- dinitrogenase reductase (also called component II) and the MoFe protein-dinitrogenase (also called component I). MoFe stands for molybdenum-iron, an essential cofactor in dinitrogenase. Some N_2 fixers have an alternative dinitrogenase that contains vanadium (V) instead of Mo. But this type of dinitrogenase is synthesized only in the absence of available Mo. The two enzymes in the complex work in tandem: dinitrogenase reductase reduces dinitrogenase, and dinitrogenase reduces N_2. Both enzymes are required for fixation to occur.

An equation for N_2 can be written as follows: Mg^{2+} is a required cofactor:

$$N_2 + 6e^- + 6H^+ + 12ATP + Mg^{2+} \longrightarrow 2NH_3 + 12ADP + 12Pi + Mg^{2+}$$

Electrons come from organic and inorganic sources, flow through several electron carriers, flow through nitrogenase, and ultimately reduce N_2:

$$e^- \longrightarrow \text{Ferredoxin} \longrightarrow \text{Flavadoxin} \longrightarrow \text{Fe protein} \longrightarrow \text{MoFe protein} \longrightarrow N_2$$

Each electron transferred requires two ATP. Unfortunately, nitrogenase is not perfectly efficient. It also reduces other compounds such as H^+, N_2O, N_3^-, and CN^-. Hydrogen evolution is a general by-product of N_2 fixation, particularly to low electron flux:

$$2H^+ + 2e^- + 4ATP \longrightarrow H_2 + 4ADP + 4Pi$$

The overall equation for nitrogenase - catalyzed N_2 fixation is therefore:

$N_2 + 16ATP + 8e^- + 8H + Mg^{2+} \longrightarrow 2NH_3 + H_2 + 16\ ADP + 16Pi + MG^{2+}$

Some organisms have an uptake hydrogenase (Hup) that allows them to reoxidize H_2 and regain part of the lost energy. In studies comparing *Rhizobium* Hup^+ and Hup^- strains, there was increased N_2 fixation and plant yield with Hup^+ strains that had the uptake hydrogenase.

Nitrogenase also catalyses the transformation of substrates such as acetylene:

Acetylene (C_2H_2) + $2H^+ + 2e^-$ $\longrightarrow$ Ethylene (C_2H_4)

This is useful because it can be used to estimate N_2 fixation rates:

$2e^- + C_2H_2 \longrightarrow C_2H_4$

$6e^- + N2 \longrightarrow 2NH_3$

Ideally, the ratio of acetylene reduced to nitrogen fixed is 3:1. But this ratio is never ideal, because it does not take into account reduction of alternative substrates such as H^+. So, N_2 fixers can be compared in terms of rates of C_2H_4 evolution, but to actually figure out how much N is being fixed, the acetylene reduction rates have to be calibrated with ^{15}N for each organism.

What is the cost of N_2 fixation? First, there is the synthesis of nitrogenase and hydrogenase (if the organism has an uptake hydrogenase). Second, there is a cost of NH_3 assimilation, which typically begins with glutamine synthetase, an ATP-requiring process. Finally, there is the lost ATP to fuel nitrogenase that could otherwise be used for growth. Figure 15 summarizes the pathway of biological N_2 fixation.

Use of Nitrogen Fixing Organisms as Biofertilizers

Biofertilizers are preparations containing microorganisms in sufficient numbers helping plant growth and nutrition. Biofertilizers are widely accepted as low cost supplements to chemical fertilizers and have no deleterious effects either on soil health or environment. Biofertilizers, which can supply nitrogen and phosphorus, two major plant nutrients, have been developed. Nitrogen present abundantly (78% by volume) in the atmosphere is fixed by several bacteria living freely in soil or in symbiosis with plants.

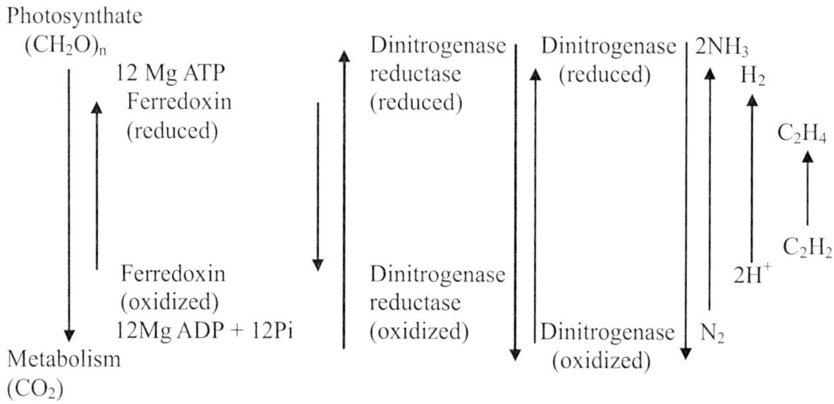

Fig. 15 : The pathway of electrons through nitrogenase during nitrogen fixation.

Rhizobium

Rhizobia isolated from root nodules are screened for efficiency through plant tubes, Leonard jars, pot experiments, microplot experiments and finally multilocational field trials. Efficient strains, thus, selected are mass multiplied in broth, mixed with carrier material like lignite, packed in polythene bags and stored either in cold room or at room temperature. The shelf life of the inoculum is usually 6 months and it should contain a viable rhizobial count of 10^8 to 10^9 cells/g lignite. Legume seeds are treated with an adhesive like sucrose or jaggery solution and then rhizobial inoculum is added and mixed thoroughly. About 450-500 g inoculum is required for treating seeds to be sown in a hectare. Seed treatment is usually done either in the early morning or late evening, and air dried before sowing. Seeds, thus treated contain 10^5 to 10^6 rhizobia/seed. Rhizobia are compatible with fungicides like captan, cerasan and insecticide like malathion. For sowing in very acidic soils, lime pelleting at the rate of 9 kg lime / ha can be done over and above rhizobial treatment. Several experiments conducted in India have shown nearly 50% nitrogenous fertilizer can be saved through inoculation with efficient rhizobia. Studies conducted at different parts of India brought out increased growth and grain yield due to *Rhizobium* inoculation (Plate 7; Table 6).

Plate 7 : Response of soybean to rhizobial inoculation
U – uninoculated, I – Inoculated

Table 6 : Quantity of nitrogen fixed and percentage increase in yield of legumes due to *Rhizobium* inoculation

Crop	N fixed kg N/season	Location over control (%)	Increase in grain yield
Chick pea	85-110	Dholi (Bihar)	12-25
		Delhi (U.T)	18-28
		Hisar (Haryana)	24-43
		Dohad (Gujarat)	33-67
		Sehore (M.P)	20-41
		Hyderabad (A.P)	No response
		Badnapur (Maharastra)	8-12
Cowpea	80-85	Krishi Nagar (Gujarat)	1-36
Lentil	90-100	Pantnagar (U.K.)	4-26
		Dholi (Bihar)	5-17
Mung bean	50-55	Lam (A.P)	1-17
		Dholi (Bihar)	16-49
		Delhi (U.T)	13-33
		Pantnagar (U.K.)	4-15
Pigeon pea	168	Varanasi (U.P)	4-19
		Dholi (Bihar)	25-12
		Delhi (U.T)	18-28
		Hisar (Haryana)	24-43
		Dohand (Gujarat)	33-67
		Sehore (M.P)	20-41
		Hyderabad (A.P)	No response
		Badnapur (Maharastra)	8-12
Urad bean	50-55	Pudukottai (T.N)	4-21
		Dholi (Bihar)	11-29
		Pantnagar (U.K.)	17-21

Azorhizobium caulinodans forms stem nodules in the green manure plant *Sesbania rostrata.* About 50 days before transplanting rice, field is sown with *Sesbania rostrata* treated with *A.caulinodans* and super phosphate applied at the rate of 50-75 kg/ha. The green manure plant is incorporated into field before transplanting rice.

Frankia

The actinomycete *Frankia* is known to form nodules in non-leguminous trees and shrubs like *Casuarina, Alnus, Ceonothus, Coriaria, Myrica* etc. and fix atmospheric nitrogen up to 300 kg N/ha/year.

Nitrogen fixed by *Frankia* in various non-leguminous trees and shrubs is given in table 7. *Frankia* is a very slow growing organism on culture media and thus no commercial inoculant preparation is available.

Table 7: Nitrogen fixed by *Frankia* in various non-leguminous trees and shrubs

Species	Kg N/ha/yr
Alnus crispa	115-362
Alnus glutinosa	56-130
Alnus rubra	325
Casuarina equisetifolia	58
Ceanothus sp.	60
Coriaria arborea	129-192
Dryas drummondii	12
Hippophae rhamnoides	15-179
Myrica gale	9

Azotobacter

Freeliving nitrogen fixing bacteria *Azotobacter* grown on Ashby's or Waksman's No.77 or Jensen's medium is mixed with the carrier material and used as described earlier for *Rhizobium. Azotobacter* can be used for cereals, vegetables, mulberry etc. Nearly 25-50% of N fertilizer application can be reduced through *Azotobacter* inoculation (Tables 8 & 9).

Table 8: Effect of *Azotobacter* inoculation at different levels of N on sugarcane yield

N level	Cane yield (t/ha) *Azotobacter*	control
150 kg N/ha	175.58	149.61
200 kg N/ha	195.92	171.36
250 kg N/ha	196.03	177.39

Azotobacter 2 kg /ha in two equal splits applied at 30 and 60 days after planting mixed with 500 kg /ha FYM.

Table 9: Effect of *Azotobacter* inoculation on yield and leaf nutrient content of mulberry

	Recommended Fertilizer	**Biofertilizer**
Leaf yield/ha/yr (Kg)	46,894	46,902
Leaf N (%)	3.297	3.306
Leaf P (%)	0.506	0.541
Leaf K (%)	1.69	1.63

Biofertilizer: 20 kg *Azotobacter* / ha/ yr in 5 split doses @ 4 kg each time mixed with 200 kg of FYM applied between rows in furrows

Recommended fertilizer for mulberry: 300 kg N + 120 kg P + 120 kg K per ha. P & K applied as per recommendation, N-150 kg/ha/yr applied 15 days before applying biofertilizer.

Azospirillum and *Gluconacetobacter diazotrophicus*

These bacteria live in associative symbiosis with the host. *Azospirillum* can be multiplied on nitrogen free bromothymol blue medium, mixed with the carrier material and used as described earlier for *Azotobacter*. Recent studies in India have shown up to 11% increase in yield of cereals like rice, wheat, sorghum, maize and pearl millet. Yield response of cereals to *Azospirillum* inoculation at different levels of nitrogen application in the field is given in table 10. A new bacterium *Herbasprillum*, taxonomically closely related to *Azospirillum* was isolated from Brazil. It was found to be associated with grasses and fix atmospheric nitrogen. This bacterium was also isolated from 17 forage grasses in and around Bangalore and Chennai. *Acetobacter diazotrophicus* is a saccharophilic bacterium and thus is associated with plants like sugarcane, sweet potato and sweet sorghum. These bacteria can be multiplied in nitrogen free malate medium and mixed with the carrier material and used for inoculating sugarcane.

Table 10 : Yield response of cereals to *Azospirillum* inoculation at different levels of nitrogen application in the field.

Crop	Nitrogen applied (Kg/ha)	Grain yield (t/ha) control	Inoculated	Increase (t/ha)
Rice	0	2.93	3.08	0.15
	30	3.52	3.98	0.46
	45	3.93	4.54	0.61
	60	4.45	4.85	0.40
Wheat	0	3.78	3.90	0.12
	40	4.54	4.58	0.04
	80	4.94	5.75	0.81
	120	4.95	6.29	1.34
Pearl Millet	0	1.79	1.91	0.12
	20	2.43	2.48	0.05
	100	2.62	2.79	0.17
Sorghum	0	3.26	3.92	0.66
	20	3.92	4.79	0.87
	40	4.46	5.38	0.92
	60	5.12	5.82	0.70
Maize	0	3.43	3.90	0.47
	33	3.95	5.81	1.86
	66	3.98	7.20	3.22
	100	4.15	7.27	3.12

Cyanobacteria

Cyanobacteria (blue green algae) like *Nostoc, Anabaena, Tolypothrix* etc. can be multiplied by the farmers in plots of 20 m x 1 m x 22 cm lined with polythene sheet and pH neutralized by adding lime. Starter culture is inoculated in the plots with water. A thick layer of algal growth, which develops in 15 days, is allowed to dry, flakes collected and used as inoculum. To control predators and insects carbofuran or ekilux at the rate of 125 g/plot can be used. From one plot of 20 m^2 up to 10 kg algal inoculum can be obtained. This inoculum can be stored for many years.

In the main field, algal inoculum is broadcast over standing water at the rate of 10 kg/ha, one week after transplanting rice seedlings. Experiments conducted have shown nearly 10% increase in grain yield due to algal inoculation and an addition of 25-30 kg N/ha/season. Field response of rice to cyanobacterial inoculation at different levels of fertilizer application is given in table 11.

Table 11: Average rice yield (q / ha) due to cyanobacterial (10 kg / ha) inoculation on – farm trials conducted in two major rice growing states.

Treatment	Madhya Pradesh (56 trials)		Tamil Nadu (118 trials)	
	Grain	% Increase	Grain	% Increase
0 N: 40 P: 15 K	24.1			
0 N: 40 P: 15 K + CB	28.2	17.0		
100 N: 50 P: 50 K			51.1	
100 N: 50 P: 50 K + CB			54.8	7.2

CB = Cyanobacteria

Azolla

Azolla is a water fern harbouring nitrogen fixing alga *Anabaena azollae* in the leaf (plate 8). *Azolla* is grown in 1 m^2 plots, 3 weeks before transplanting rice. Inoculation is done at the rate of 100 g/m^2 (*i.e.* 0.1 t/ha). After 15 to 20 days, *Azolla* multiplies 100 folds, resulting in a yield of 10-15 t/ha. Furadon (0.02%) is added to prevent insect damage. *Azolla,* thus multiplied, is added to the main field at the rate of 10 tonnes fresh material/ha and incorporated into soil before transplanting rice. The other method includes growing *Azolla* along with standing crop of rice as dual crop. The incorporation method is better. About 30-50 kg N/ha/season is added by *Azolla*. It also adds organic matter and thus improves the physico-chemical properties of soil. Field experiments have shown that nearly 50% N fertilizer can be saved through *Azolla* application. Increase in rice yield because of *Azolla* inoculation is given in table 12.

Plate 8 : *Azolla*

Table 12: Effect of *Azolla microphylla* inoculation and urea on the grain yield of rice.

Treatment	Variety IR 64		Variety IR 50	
	Grain yield (kg/ha)	% Increase over control	Grain yield (kg/ha)	% Increase over control
Control	2735	-	3071	-
Azolla microphylla	2905	6.22	3172	3.29
Urea (100 kg N/ha)	3085	12.80	3212	4.59
A.microphylla + Urea	3435	25.59	3374	9.87
CD (P=0.05)	NS		214	

References

Bagyaraj, D.J., Krishnaraj, P.U. and Khanuja, S.P.S. 2000. Mineral phosphate solubilization: Agronomic implications, mechanism and molecular genetics. *Proceedings of Indian National Science Academy. (PINSA)*. B, 66: 69-82.

Baldani, V.L.D., Baldani, J.I. and Dobereiner, J. 2000. Inoculation of rice plants with the endophytic diazotrophs *Herbaspirillum seropedicae* and *Burkholderia* spp. *Biological Proceedings of the Royal Society of London,* 127 B: 129-139.

Benson, D.R. and Clawson, M.L. 2000. Evolution of the actinorhizal plant symbioses. In: *Prokaryotic Nitrogen Fixation: A Model System for Analysis of Biological Process,*(ed. E.W. Triplett), Horizon Scientific Press, Wymondham, UK, pp.207-224.

Benson, D.R. and Silvester, W.B. 1993. Biology of *Frankia* strains, actinomycete symbionts of actinorhizal plants. *Microbial Reviews.* 57:293-319.

Boddey, R.M., Urquiaga, S., Alves, B.J.R. and Reis, V. 2003. Endophytic nitrogen fixation in sugarcane: present knowledge and future application. *Plant and Soil,* 252:139-149.

Callaham, D., Del Tredici, P. and Torrey, J.G. 1978. Isolation and cultivation *in vitro* of the actinomycetes causing root nodulation in *Comptonia. Science,* 199: 899-902.

Carrapico, F., Teixeira, G. and Diniz, M.A. 2000. *Azolla* as a biofertiliser in Africa - A challenge for the future. *Revista de Ciencias Agrarias,* 23: 120-138.

Chovance, P., Hovoraka, O. and Novak, K. 2008. Visualization of Symbiotic tissue in intact root nodules of *Vicia tetrasperma* using GFP-marked *R. leguminosarum* bv. *viciae. Folia Microbiologica,* 53:139-146.

Dar, G.H. 2000. *Soil Microbiology and Biochemistry*. NIPA Pub., New Delhi, India.

Dazzo, F.B., Yanke, W.E. and Brill, W.J. 1978. Trifolin – A *Rhizobium* recognition protein from white clover. *Biochemical Biophysical Acta,* 539: 276-286.

Deaker, R., Roughley, R.J. and Kennedy, I.R. 2004. Legume seed inoculant technology - A review. *Soil Biology and Biochemistry,* 36: 1275-1288.

Dobereiner, J. 1997. Biological nitrogen fixation in the tropics: Social and economic contributions, *Soil Biology and Biochemistry*, 29: 771-774.

Dobereiner, J., Burris, R.H. and Hollaender, A. (eds.) 1978. *Limitations and Potential for Biological Nitrogen Fixation*. Plenum Press, New York, USA.

Dommergues, Y.R. 1997. Contribution of actinorhizal plants to tropical soil productivity and rehabilitation. *Soil Biology and Biochemistry,* 29:931-941.

Graham, P.H. and Vance, C.P. 2000. Nitrogen fixation in perspective: an overview of research and extension needs. *Field Crops Research.,* 65:93-106.

Hirsch, A.N., Lum, M.R. and Downie, J.A. 2001. What makes rhizobia-legume symbiosis so special ? *Plant Physiology,* 127:1481-1492.

Kannaiyan, S., Kumar, K. and Gopalaswamy, G. 2001. Stem nodulating *Sesbania rostrata* as a bioresource for rice production, induction of nodulation and endophytic N_2 fixation in rice. *Madras Agricultural Journal,* 88:547-560.

Kennedy, I.R. and Islam, N. 2001. The current and potential contribution of asymbiotic nitrogen fixation to nitrogen requirements on farms: a review. *Australian Journal of Experimental Agriculture,* 41:447-457.

Lewin, A., Rosenberg, C., Meyer, H., Wong, C.H., Nelson, L., Manen, J.F., Stanley, J., Dowling, D.N., Denarie, J. and Broughton, W.J. 1987. Multiple host-specificity loci of the broad host - range *Rhizobium* sp. NGR234 selected using the widely compatible legume *Vigna unguiculata. Plant Molecular Biology,* 8: 447-459.

Mishra, U. and Pabbi, S. 2004. Cyanobacteria - A potential biofertilizer for rice. *Resonance*, June: 6-10.

Moreira, F.M.S. 2008. Nitrogen-fixing Lequminosae-nodulating Bacteria.In: *A Hand Book of Tropical Soil Biology – Sampling and Characterization of Below-ground Biodiversity* (eds. F.M.S., Moreira, E.J., Huising., and D.E., Bignell), Earthscan Pub., London. pp. 107-130.

Muthukumarasamy, R., Revathi, G., Seshadri, S. and Lakshminarasimhan, C. 2002. *Gluconacetobacter diazotrophicus* (syn. *Acetobacter diazotrophicus*) a promising diazotrophic endophyte in tropics. *Current Science,* 83:137-45.

Nautiyal, C.S. 2000. Plant beneficial rhizosphere competent bacteria. *Proceedings of the National Academy of Sciences* (India), 70 (B):107-123.

Oakely, B., North, M., Franklin, J.F., Hedlund, B.P. and Staley, J.T. 2004. Diversity and distribution of *Frankia* strains symbiotic with *Ceanothus* in California. *Applied Environmental Microbiology,* 70:6444-6452.

Olivares, F.L., Baldani, V.L.D., Reis, V.M., Baldani, J.I. and Dobereiner, J. 1996. Occurrence of the endophytic diazotrophs *Herbaspirillum* spp. in root, stem, and leaves, predominantly of Gramineae. *Biology and Fertility of Soils,* 21:197-200.

Paul, S. and Verma, O.P. 2005. *Azotobacter*- Bioinoculants for sustainable agriculture. In: *Advances in Microbiology*. (ed. B.D.Kaushik), Indian Agricultural Research Institute Pub., New Delhi, India. pp. 135-146.

Quispel, A. 1991. A critical evaluation of the prospects for Nitrogen fixation with non - legumes. *Plant and Soil*, 137:1-11.

Rai, A.N. 1990. Cyanobacteria in symbiosis.In: *Handbook of Symbiotic Cyanobacteria* (ed. A.N. Rai), CRC Press, Boca Raton, USA.

Saleena, L.M., Rangarajan, S. and Nair, S. 2002. Diversity of *Azospirillum* strains isolated from rice plants in saline and non-saline coastal agricultural ecosystem. *Microbial Ecology*, 44:271-277.

Saxena, A.K. 2005. *Rhizobium* inoculants: Research and development. In: *Advances in Microbiology*. (ed. B.D. Kaushik), Indian Agricultural Research Institute Pub., New Delhi, India. pp. 73-92.

Schwintzer, C.R. and Tjepkema, J.D. (eds.) 1990. *The Biology of Frankia and Actinorhizal Plants*. Academic Press Inc., San Diego, USA.

Singh, P.K. and Singh, D.P. 1996. *Azolla-Anabanea* symbiosis. In: *Biotechnological Approaches in Soil Microorganisms for Sustainable Crop Production*. (ed. K.R. Dadarwal), Scientific Publishers, Jodhpur, India. pp. 93-107.

Subba Rao, N.S. and Dommergues, Y.R. (eds.) 1998. *Microbial Interaction in Agriculture and Forestry. Vol.1*. Oxford & IBH Pub. Co., New Delhi, India.

Subba Rao, N.S. and Rodriguez-Barueco, C. (eds.) 1993. *Symbioses in Nitrogen Fixing Trees*. Oxford & IBH Pub. Co., New Delhi, India.

Thajuddin, N. and Subramanian, G. 2005. Cyanobacterial biodiversity and potential applications in biotechnology. *Current Science*, 89:47-57.

Tilak, K.V.B.R. 2005. Progress on *Azospirillum* as biofertilizer. In: *Advances in Microbiology*. (ed. B.D. Kaushik), Indian Agricultural Research Institute Pub., New Delhi, India. pp. 93-102.

Tilak, K.V.B.R., Pal, K.K. and Dey, R. 2010. *Microbes for Sustainable Agriculture*. I.K. International Pub. House Pvt. Ltd., New Delhi, India.

Tilak, K.V.B.R. and Saxena, A.K. 2000. *Azospirillum*- its impact on crop production. In: Recent *Advances in Biofertilizer Technology* (eds. A.K. Yadav, M.R. Motsara and S. Ray Chaudhuri), Society for Promotion & Utilization of Resources and Technology, New Delhi, pp. 190-206.

Venkataraman, G.S. 1972. *Algal Biofertilizers and Rice Cultivation*. Today and Tomorrow's Printers and Publishers, New Delhi, India.

Vincent, J.M. 1970. *A Manual for the Practical Study of Root-Nodule Bacteria*. Blackwell Scientific Publications, Oxford, UK.

Chapter-4

Microorganisms Modulating Phosphorus Nutrition of Plants

Phosphorus is one of the essential nutrients for plant growth. It is an integral part of the cellular activities of the living organisms. It has a defined role in plant metabolism such as cell division, development, photosynthesis, break down of sugar, nutrient transport within the plant, transfer of genetic characteristics from one generation to another and regulation of metabolic pathways. Phosphorus is a frequently limiting macronutrient next only to nitrogen for plant growth and makes up about 0.2% of plant dry weight. The plants obtain their P requirement from the soil pool. It occurs in soil as inorganic phosphates, produced by weathering of parent rock or as organic phosphate derived from decayed plant, animal, or microorganisms.

Considerable interest in the past has gone into the development of rock phosphate directly as fertilizers, mainly because nearly 60% of the 145 million tonnes of rock phosphate deposits in India are considered to be very inefficient ore for manufacture of phosphate fertilizer through wet processing. Mussorie rock phosphate, the most reactive of different phosphates available in India has been used in acid soils .The mineral phosphate solubilizing microbes, which acidify the substrate and possibly the rhizosphere, have been used in unison with rock phosphate so that the phosphorus in rock phosphate can be made available in soil for plant uptake.

Phosphorus in Soil

Soil is a dynamic system and is an ecological niche of constant biological activity, influenced to a great extent by the chemical nature of its parent material and the plant growth it supports. The Pi available for biosynthetic purposes will depend not only on the total amount of phosphorus in the environment but also on its solubility, which in turn is dictated by several chemical reactions and biological interactions in the soil. The plants have to pick up their Pi requirements from soil and the soil bacteria too must accumulate Pi from the soil for their biochemical activities. Recycling of phosphate in the soil is depicted in fig. 16. The diverse soil P forms can be generally categorized as soil solution P, insoluble organic and insoluble inorganic P. In any agriculture system, not all the applied phosphorus is available to the crop.

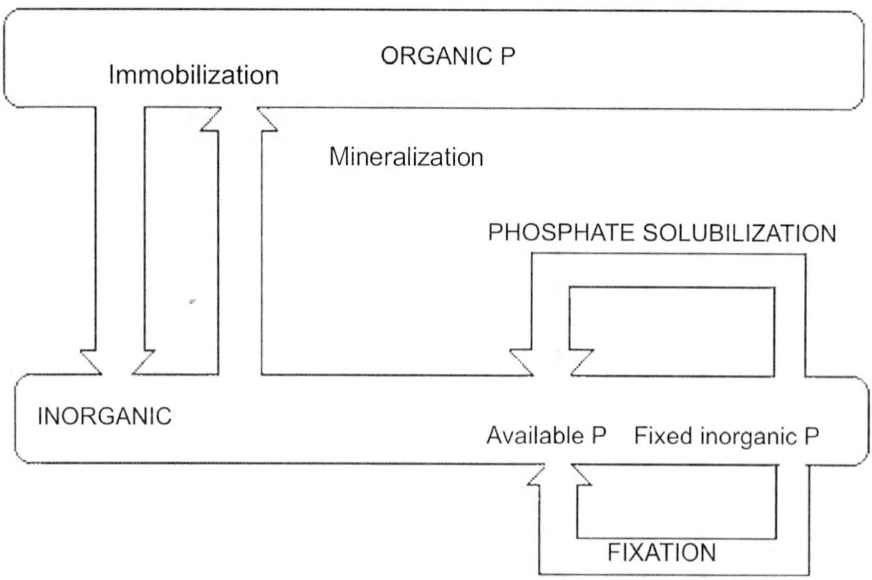

Fig. 16 : Phosphorus channel in soil

In soil the two reactions, fixation and immobilization convert applied phosphorus into forms unavailable for the plant. More than 70-90 % of the applied phosphatic fertilizers get fixed in the soil rendering them unavailable for plant uptake. The decomposing roots also release phosphorus because of autolysis, directly into the soil solution as inorganic orthophosphates. The ultimate result of this is also fixation. Great percentages of soil P are also converted to organic forms of which inositol hexa phosphate is usually a major component, and thus get immobilized and not available for plant growth. Hence, the concept of conversion of inorganic unavailable phosphate into available forms *viz.*, $H_2PO_4^-$ and HPO_4^{-2} for plant uptake, a phenomenon referred to as mineral phosphate solubilization. The form in which Pi exists also changes according to the soil pH. Below pH 6 most Pi will be present as monovalent H_2PO_4 species. The plant uptake is also high at the pH range of 5-6, which indicates that P is primarily taken up as monovalent form. The average orthophosphate concentration in the soil solution of around 10^{-6} M is near the limit at which plants can absorb adequate phosphate. However, critical (threshold) concentrations vary for different plant species. Phosphorus enters the plant root hairs, root tips and the outer most layers of the root cells. Once inside the roots, the inorganic P is stored in the root or transported into the upper portions of the plant. There through various chemical reactions it is incorporated into organic compounds.

Phosphorus fertilizers are regularly applied to get maximum yields. But as a result of chemical reactions in the soil, these get fixed in the soil resulting only part of it being readily available over the crop period, necessitating, fresh additions. Continuous application of these fertilizers will result in increased concentration of total phosphorus in the soil over time, resulting in large reserves of fixed P. This is an important sink that needs to be tapped for phosphorus nutrition. Less than 10% of soil P enters the plant-animal cycle. Consequently P deficiency is a wide spread problem and P fertilizers are almost universally required for crop production.

Amongst the various bioameliorants available, the mineral phosphate solubilizing microorganisms play an important role in recycling phosphorus in the soil.

Mineral Phosphate Solubilizing Microorganisms

The fixed phosphorus in the soil can be solubilized by certain microbes, which have the capacity to convert inorganic unavailable P forms *viz.* varisite [$Al(OH)_2H_2PO_4$], strengite [$Fe(OH)_2H_2PO_4$], flourapatite [$Ca_{10}PO_4F_2$], hydroxyapatite [$Ca_5(PO_4)_3OH$] and tricalcium phosphate $Ca_3(PO_4)$, to forms available for plant growth. The mineral forms of iron and aluminium phosphates are predominant in acidic soils, while calcium phosphates predominate in neutral to alkaline soils. These unavailable forms are converted to primary orthophosphate (H_2PO_4) and secondary orthophosphate ($H_2PO_4^{-2}$), which are available for plant growth. Such microbes are said to possess mineral phosphate solubilizing (MPS) ability. The MPS microbes, which include bacteria, fungi and actinomycetes, are ubiquitous and their number and type varies with soil, climate and vegetation. Experiments with the MPS microbes have proved their usefulness on crop growth and yield. Several reviews are available on the phosphate solubilizing microorganisms.

Traditionally the MPS microbes have been isolated on media in which insoluble phosphates like dicalcium phosphate or tricalcium phosphate is used as a sole source of phosphorus. Indicator medium containing bromothymolblue (BTB) also has been developed to enhance chances of picking up efficient bacteria. A clear zone around the colony indicates the ability of release of Pi from the precipitate of insoluble phosphate. More vigorously, the cultures are tested for their Pi release in liquid medium containing insoluble phosphates. The release of Pi is tested spectrophotometrically. Such analysis of different MPS microbes showed that the P solubilization varied from 4.36 - 59.4%.

Mechanism of Phosphate Solubilization

A number of theories have been advanced to explain the mechanisms of P-solubilization by microorganisms. Important among them are: (a) Acid production theory (b) Proton and enzyme theory.

Acid production theory

According to this theory, the process of phosphate solubilization by PSM is due to the production of organic acids, which is accompanied by the acidification of the medium. Many workers recorded a decrease in the pH of the filtrate from an initial value of 7.0 to a final value of 2.0.The analysis of culture filtrates of MPS microbes has shown the presence of a number of organic acids such as malic, glyoxalic, succinic, fumaric, tartaric, alpha keto butyric, oxalic, citric, 2-ketogluconic and gluconic acids. The organic acids released in the culture filtrate react with the insoluble phosphate. The amount of soluble phosphate released depends on the strength and type of acid. Besides organic acids, inorganic acids such as nitric and sulphuric acids are also produced by the nitrifying bacteria and *Thiobacillus* sp. during the oxidation of nitrogenous or inorganic compounds of sulphur, which react with calcium phosphate and convert them into soluble forms.

Proton and enzyme theory

Esterase type enzymes are known to be involved in liberating phosphorus from organic phosphatic compounds. MPS micro-organisms are also known to produce phosphatase enzyme along with acids, which cause solubilization of P. Protons H^+ released originating from NH_4 assimilation and respiration could bring about solubilization of mineral phosphate as protons bring in a rapid reduction in pH and increase Pi in the medium. It is also believed that gluconic acid produced by microorganisms lead to release of protons that solubilized insoluble phosphates.

The mechanism of MPS activity is therefore due to release of organic acids, which could also contribute protons to extensively acidify the medium, ultimately causing the dissolution of mineral phosphate. Beside these two mechanisms production of siderophores, humic substances and CO_2 are also correlated with P solubilization.

Siderophores

These iron chelating compounds produced by MPS microbes and certain PGPR bacteria chelate iron from iron phosphates and thereby increases P availability in acid soils.

Humic substances

Humic and fulvic acids with carboxylic, phenolic, hydroxylic or phenol hydroxyl moieties release P from various P sources such as compost, rock phosphates etc. due to formation of stable complexes with calcium, iron and aluminium.

Carbon dioxide

Rhizospheric microbes during respiration release CO_2, that is converted to carbonic acid decreasing pH of microenvironment that helps in P solubilization.

Phosphate mineralization is attributed to presence of two specific genes in MPS^+ phenotype *viz.* the gluconic acid dehydrogenase (GDH) that converts glucose to gluconate and the cofactor pyrroloquinoline quinone (PQQ) synthase genes (forms a peculiar quinoprotein required for MPS^+ phenotype).

Use of P Solubilizing Microorganisms as Biofertilizers

Tropical soils are deficient in phosphorus. Further, most of them fix P and thus makes it unavailable for plant growth. It is estimated that in most tropical soils, 75% of super phosphate applied is fixed and only 25% is available for plant growth. There are some fungi and bacteria like *Bacillus polymyxa, B. megaterium, Pseudomonas striata, Aspergillus awamori,* and *Pencillium digitatum* which can solubilize unavailable form of P to available form (Plate 9). India has 260 mt of rock phosphate deposits. The cheaper sources of rock phosphate like Mussorie rock phosphate, Udaipur rock phosphate etc., available in our country can be used along with phosphate solubilizing microorganisms. This will save considerable amount of foreign exchange, as the raw material for the manufacture of superphosphate is imported.

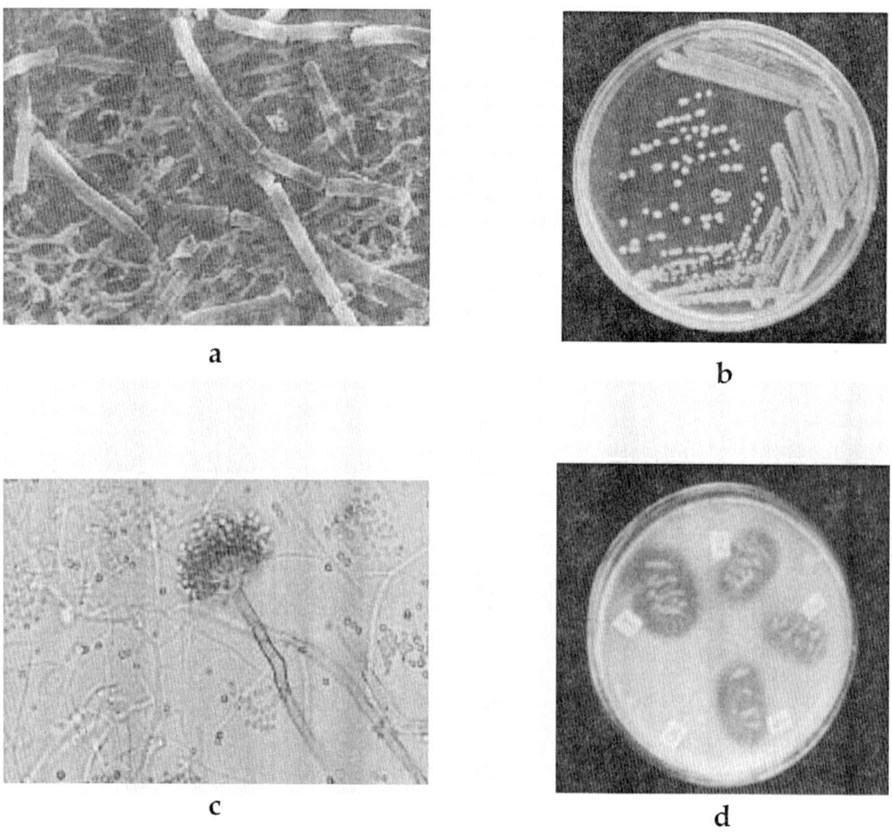

Plate 9 : a. *Bacillus megaterium* cells, b. Plate showing growth of *B. megaterium* c. *Aspergillus awamori*, d. Plate showing P solubilizing zones

The P solubilizing bacteria or fungi can be mass multiplied on Pikovskaya's broth and mixed with the carrier material used. Several studies conducted by scientists brought out that by inoculation with phosphate solubilizing microorganisms along with rock phosphate, it is possible to produce the same amount of yield as with superphosphate. Yield increase because of inoculation with P solubilizing organisms has been reported in different crops (Table 13).

Table 13 : Field response of crop plants to inoculation with phosphate solubilizing microorganisms

Crop	Yield (kg/ha) Uninoculated	Inoculated	% Increase
Rice	1,937	2,161	11.50
	2,050	2,300	12.20
Wheat	4,100	4,811	17.30
	4,786	5,584	16.60
Chickpea	2,370	2,920	23.20
Pea	2,169	2,717	25.20
Soybean	1,050	1,786	70.00
Potato	23,330	37,330	60.00

Mycorrhizal Fungi

German Botanist Albert Bernhard Frank in 1885 introduced the Greek word 'mycorrhiza' into scientific terminology, which literally means 'fungus root'. Mycorrhiza results from a mutualistic symbiosis between roots of higher plants and certain fungi. Though the word was introduced in 1885, mycorrhizae itself, of course, are millions of years older, for fossil mycorrhizae have been found in carbonaceous deposits. In natural ecosystems, it is exceptional for a plant not to possess a mycorrhizal root system. So it could be said that mycorrhizal association is very common or almost universal phenomenon in plant kingdom.

There are different types of mycorrhizae. The most common are: 1. Ectomycorrhizae 2. Ericoid mycorrhizae 3. Orchidaceous mycorrhizae and 4. Arbuscular mycorrhizae.

Ectomycorrhiza

Ectomycorrhizae are most common among forest and ornamental tree species in the families Pinaceae, Salicaceae, Betulaceae, Fagaceae and Tiliaceae as well as in some members of Rosaceae, Leguminaceae, Myrtaceae and Juglandaceae. Numerous fungi have been identified as forming ectomycorrhizae. Worldwide, over 5,000 species of fungi can form ectomycorrhizae on some 2,000 species of woody plants. Among the basidiomycetous fungi, species of Hymenomycetes, in the genera *Boletus, Cortinarius, Suillus, Russula, Hebeloma, Tricholoma, Laccaria* and *Lactarius* and species of Gasteromycetes in the genera *Rhizopogon, Scleroderma, Alpova* and *Pisolithus* form ectomycorrhizae. Certain orders in the Ascomycetes such as Eurotiales (*Cenococcum geophilum*) and Tuberales (*Tuber* sp.) also have species that form ectomycorrhizae (Plate 10). Thus, they are mostly fungi forming mushrooms, puffballs or truffles.

Plate 10 : Ectomycorrhizal fungi. a. *Boletus pinophilus*, b. *Suillus cavipes*, c. *Russula decipiens*, d. *Laccaria laccata*, e. *Pisolithus tinctorius*, f. *Cenococcum* sp.

Structurally ectomycorrhizae can be distinguished by the presence of thick fungal mantle or sheath around the feeder roots. From the mantle, fungal strands radiate outside into the soil thus enhancing root surface absorptive area. Hyphae also penetrate through the epidermis into the intercellular spaces of the cortical cells apparently replacing the middle lamella and forming an interconnecting network known as 'Hartig net'. Ectomycorrhizal roots are generally short, swollen, dichotomously branched and with distinctive colours of white, black, orange, yellow or olive green. Some fungi are apparently host specific, others have broad host ranges and form mycorrhizae with members of numerous tree genera in diverse families. For example, *Cenococcum* has been reported to have widest host range but *Boletus elegans* is very host specific and forms association with only *Larix* sp.

The ectomycorrhizal fungi can be isolated in the vegetative form although the identification of such fungi becomes difficult since the reproductive bodies are not readily formed in culture media. Fruiting bodies can, however, be seen on the soil surface near the trees from which fungal cultures can easily be isolated. These fungi grow slowly in culture and require special nutrients such as thiamine and simple amino acids. Modified Melin Norkran's medium is commonly used for culturing ectomycorrhizal fungi. Optimum pH for the growth of the fungi is 5 - 5.5 and optimum temperature is 25 - 30°C. The fungus obtains photosynthates from the host and provides certain nutrients to the host.

Mycorrhizal roots lack root hairs. The fungal sheaths together with the hyphae extending to the soil absorb nutrients. The Hartig net acts as a liaison tissue between the fungal sheath and the host cells. The ectomycorrhizal habit increases the surface area of the root system and hence affords better intake of nutrients such as nitrogen, phosphorus and potassium from the surrounding soil. It is likely that the sheath acts as phosphate reservoir and releases the nutrient during certain deficiency conditions while under normal conditions, a steady uptake of phosphate to the plants is maintained by the fungal hyphae. Ectomycorrhizal fungi can also solubilize unavailable forms of P to available forms. They are also cellulose decomposers and thus help in the decomposition of organic matter. They also help in water uptake and produce growth-promoting substances like IAA, cytokinins, ethylene in considerable amount. They also protect the root against invasion by root pathogens.

In developing countries, soil collected from the root zone of established plantations have been used as inoculum in nurseries and/ or to establish plantations in new sites. Spores of ectomycorrhizal fungi have also been used for the same purpose. Ectomycorrhizal fungi are now produced commercially in bioreactors in many countries and such preparations are available in the market to inoculate forest nurseries.

Ericoid mycorrhiza

Ericoid mycorrhizal fungi usually colonize plants belonging to the families Ericaceae, Empetraceae and Epacridaceae. These are temperate plants and they have very fine roots. The temperate soils in which they grow is usually peaty, acidic, and deficient in nitrogen and phosphorus. The ericaceous plants are either shrubs or small trees. They are commonly referred to as heath plants eg. azalea, rhododendron, blueberry, cranberry etc. The fungus, forming mycorrhizal association is *Hymenoscyphus ericae,* earlier called as *Pezizella ericeae,* which is an apothesium forming ascomycete.

Most of the ericoid plants do not produce root hairs and hence more dependent on the mycorrhizal fungus for nutrient uptake from soil. The roots of some host plants lack epidermis and have reduced root cap, which may sometimes be absent. Mycorrhizal infection is confined to the epidermis and the outermost or sole layer of the cortical cells and it never penetrates the endodermis or stele. Septate hyphae spread over the surface of the root without forming appressoria, however, there could be a slight swelling of the hypha at the point of penetration. The intracellular hypha is surrounded by host plasmalemma. Intracellular hyphae form characteristic coil that fills the cells and the remainder volume of the cell is filled with phenol containing vacuoles. The hyphae of ericoid mycorrhizal fungus is very thin and hence difficult to observe compared to other fungi. The fungus spreads along the surface of the root and produces branches, which infect the cortical cells. A spread from cell to cell is infrequent. In one-centimeter root length up to 2,000 infection points have been observed. The fungus can be cultured on 0.5 % malt agar. The optimum temperature for growth is between 20 and 30 ^{0}C and the optimum pH is between 6 and 7. A wide range of carbon compounds supported growth of the fungus except cellulose. The fungus produces the enzyme pectinases and showed no ability to degrade humic and fulvic acids.

Most ericaceous species characteristically grow on nutrient poor, acid soil where ammonium predominates over nitrate. Ammonium ions are relatively immobile in soil. Thus, ericoid mycorrhizal fungi help in the uptake of both N and P. The mycorrhizal plants can also take amino acids from soil as a source of nitrogen. Plants with ericoid mycorrhizal association showed tolerance to heavy metals like zinc, copper, aluminium etc. In an experiment, the shoots of the non-mycorrhizal plants contained higher metal levels than the mycorrhizal plants. In mycorrhizal plants, the heavy metals were stored in the interfacial matrix between the fungus and the host in the root. This suggests the possible use of ericoid mycorrhizal association for the revegetation of sites polluted with heavy metals.

Inoculation with the ericoid mycorrhizal fungus made the blueberry cultivation possible in New Zealand, as these soils originally did not harbour this fungus. Now New Zealand exports blueberry to several countries. Further the fungus has been mass produced commercially and available in the market. In India, azaleas and rhododendrons are grown in Himachal Pradesh and North Eastern regions. The possibility of using the ericoid mycorrhizal fungus for enhancing the productivity of these plants can be an area for future research.

Orchid mycorrhiza

Orchids belong to the family Orchidaceae. This family has nearly 30,000 species. Some of the orchids are terrestrial and some of them are epiphytes. Some of the orchids are achlorophyllous. Orchids are remarkable for two characters: firstly, their seeds are extremely small, the largest being about 14 μg and secondly, they are all mycorrhizal, living throughout life in association with the fungi. Orchid seeds contain very limited reserves in the form of starch or lipid. At the time of germination seeds absorb water, swell slightly and the seed coat breaks exposing the epidermal hair and this structure is referred to as the 'protocorm'. The protocorm has to be infected by the mycorrhizal fungus to develop into a plant. Protocorms wait up to 6 months to be infected by the mycorrhizal fungus. If not infected by the fungus it dies. Orchids are thus obligatorily dependent on mycorrhizal fungi.

The fungi involved were initially identified as *Rhizoctonia solani*, *R.repens* and other *Rhizoctonia* spp. Later their perfect stage was discovered and they belonged to the genera *Thanatephorus*, *Ceratobasidium*, *Sebacina* and *Tulasnella*. Extensive studies have been made in the orchid mycorrhizal fungus *Tulasnella calospora*. This fungus can be cultured in the laboratory. If the host is not available, it can survive as a saprophyte.

Infection of the host by fungus

There are two stages of infection in the life cycle of many orchids (i) the primary infection of the germinating protocorm and (ii) the reinfection of the new roots of the adult or annual roots. Many terrestrial orchids shed the roots annually, and the root system is renewed annually. Here infections occur rapidly in the new root, arising from the propagules in the soil or from the dead root system of the previous year. In most of the orchids, roots develop from a tuber or bulb. The tuber is never colonized by the mycorrhizal fungus because of high concentration of phytoalexins inhibitory to the fungus.

When the fungal hypha penetrates the host cell, invagination of the plasmalemma of the host surrounds the fungal hypha. The fungus forms a typical coiled structure within the cell called as 'peleton'. Only one peleton is formed per cell. The electron microscopic studies showed that the fungal cell wall has two layers (i) outer electron less dense layer (ii) inner electron dense layer. The outer layer is comparable to the interfacial matrix in AM association. Peleton is the site of exchange of nutrients between the fungus and the host. The life span of peleton is 20-40 hours. After this period, digestion of the peleton takes place resulting in exchange of nutrients. The digestion of peleton has been attributed to the activity of acid phosphatase coupled with lysozyme by some workers. Others feel that it is due to the activity of oxidative enzymes like catalase, ascorbic acid oxidase and polyphenol oxidase.

Because of infection the host cell, becomes physiologically more active, contains more mitochondria and 3-4 fold increase in the DNA content. Starch disappears in the infected cell. The most interesting information about orchid mycorrhizal symbiosis is that the fungus in addition to supplying mineral nutrients can also supply carbohydrates to the host under certain circumstances. This is very much in contrast to the other mycorrhizal symbiosis wherein the fungus gets the carbohydrates from the host and host gets the mineral nutrients from the fungus. Orchid mycorrhizal fungus can produce hydrolytic enzymes like cellulases, pectinases and amylases and thus can utilize different carbohydrates.

Phytoalexins are chemicals produced by the host because of infection by a pathogen. They are mostly dihydroxy phenanthrenes. In orchids, dihydroxy phenanthrenes produced are hircinol, loroglossol and orchinol. Maximum phytoalexin production takes place in the tubers of the orchids and perhaps this is the region why the fungus is not able to colonize the tuber. In the roots, the concentration of phytoalexins is such that it encourages effective symbiosis. In other plants, *Rhizoctonia* is a pathogen possibly because the phytoalexin concentration produced is low to ward off the pathogen.

Study by the author in New Zealand revealed that, there is an initial lag in the building of mycorrhizal infection in places where orchid cultivation is newly undertaken and also mycorrhizal fungi, once established spread rapidly from plant to plant with in a glass house suggesting that there is no need for artificial inoculation of orchids.

Arbuscular mycorrhiza

The common mycorrhizal association in most of the plants is the arbuscular type occurring in the majority of the agricultural crops, most shrubs, most tropical tree species and some temperate tree species. It is easier to list plant families that do not form arbuscular mycorrhizae (AM) than to list those that do. Families not forming AM include Pinaceae, Betulaceae, Orchidaceae, Fumariaceae, Commelinaceae, Urticaceae and Ericaceae. Families that rarely form AM include the Brassicaceae, Chenopodiaceae, Polygonaceae and Cyperaceae. Families that form both ectomycorrhizae and AM include Juglandiaceae, Tiliaceae, Myrtaceae, Salicaceae, Fagaceae and Caesalpiniaceae. In addition to the wide spread distribution of AM throughout the plant kingdom the association is geographically ubiquitous and occurs in plants growing in arctic, temperate and tropical regions. AM occurs over a broad ecological range from aquatic to desert environments.

AM fungi belong to the phylum Glomeromycota, which has a single class Glomeromycetes with 4 orders Glomerales, Diversisporales, Paraglomerales and Archaeosporales. There are 11 families, 17 genera and 228 species. The commonly occurring genera of AM fungi are *Glomus, Gigaspora, Scutellospora, Acaulospora* and *Entrophospora* (Plate 11). These fungi are obligate symbionts and have not been cultured on nutrient media. AM fungi are not host specific although evidence is growing that certain endophytes may form preferential association with certain host plants.

The AM fungi form the largest fungal resting spores known. These are often several hundred micrometers in diameter and can survive adverse soil conditions and germinate when conditions are favourable. The germ tubes die unless they encounter and successfully penetrate the host root. After forming an appressorium on the root surface, the hypha penetrates the root and ramifies in the root cortex. Branches from the longitudinally running intercellular hyphae enter cells where further branching results in highly branched hyphal structures called 'arbuscules' (Plate 12a). For many years, host digestion of arbuscules was assumed to be the nutrient transfer mechanism between the fungus and the host.

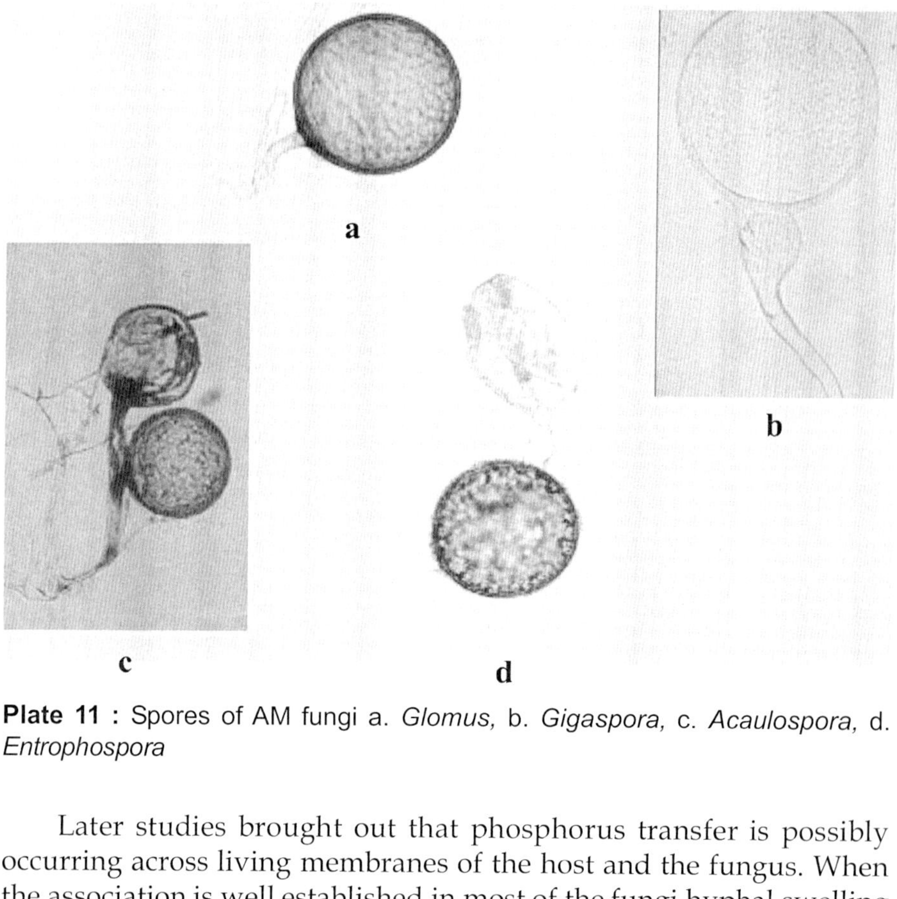

Plate 11 : Spores of AM fungi a. *Glomus,* b. *Gigaspora,* c. *Acaulospora,* d. *Entrophospora*

Later studies brought out that phosphorus transfer is possibly occurring across living membranes of the host and the fungus. When the association is well established in most of the fungi hyphal swelling termed vesicles form on the mycelium mostly inside and rarely outside the root (Plate 12b). These vesicles contain oil droplets and may function as storage organs. Mycelium outside the root form very large thick walled chlamydospores.

A root system colonized by AM fungi does not show any morphological variations from the normal root system and hence cannot be distinguished visually. In some plant species like onions, maize, clover etc. they may appear yellowish, but not in all plant species. This colour however, disappears rapidly upon exposure to light. Hence, AM fungal status of a root system can only be known through microscopic observation after staining the roots with trypan blue. The presence of vesicles and arbuscules is by definition diagnostic criterion for identifying AM fungus in a root. AM fungi can be isolated from the soil preferably from the root zone, by the wet sieving and decantation method. Single spores thus obtained can be surface sterilized by immersing in a solution containing 2% (w/v) chloramine T and 200-ppm streptomycin for 15 minutes, followed by washing 3 times in sterile water. Individual spores can be removed by a fine capillary pipette, under a dissecting microscope and placed in the collar region of a funnel filled with sterile sand and seeded with suitable host like sorghum. After the mycorrhizal synthesis in about 3-4 weeks, sorghum seedlings along with sand from the funnel can be transferred into sand:soil (1:1) mixture in a pot and maintained in the glass house as 'pot culture'. Once a pot culture of a single species is achieved, inoculum from it can be used to establish a large number of stock plants to provide inoculum for future experiments.

a

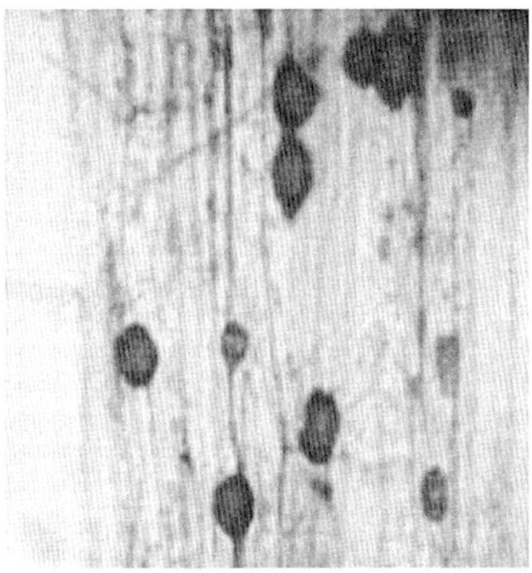

b

Plate 12 : AM fungal structures in root a. Arbuscule, b. Vesicles

Pot experiments conducted earlier in sterilized soil showed that AM inoculation can drastically improve plant growth. Since most of the natural soils usually harbour AM fungi it was felt that plants might not respond to mycorrhizal inoculation in unsterile soils. But, later investigation indicated that even in unsterile soils plants do respond to inoculation with efficient strains of AM fungi. Numerous experiments have been conducted in different parts of the world confirming this growth improvement. The increased growth of mycorrhizal plants is favoured in soils with low to moderate fertility. More specifically soils favouring mycorrhizal growth stimulation usually have one or more essential nutrients especially phosphorus in limiting concentrations.

Many workers have investigated the mechanism of improved plant growth caused by mycorrhizal inoculation. Greater soil exploration by mycorrhizal roots as a means of increasing phosphate uptake is well established. Phosphate ions are fairly immobile and a phosphate depletion zone often develops around roots in phosphate deficient soils. The hyphae are able to reach beyond this zone and directly translocate nutrients from the soil to the root cortex where transfer to the plant occurs. Experiments with ^{32}P labeled phosphate indicate that AM hyphae obtained their extra phosphate from the labile pool rather than by dissolving insoluble phosphate. The better utilization of sparingly soluble rock phosphate is explained by the hyphae making closure physical contact with the ions disassociating at the particle surface.

AM fungi have been reported to improve the uptake of other elements like Zn, Cu, K, S, Al, Mn, Mg, Fe, etc. More consistent results have been obtained with uptake of Zn and Cu. Some plants have been reported to grow better when inoculated with AM fungi than when given any amount of P fertilizer e.g. *Araucaria cunninghamii* in Australia, *Liquidambar styraciflua* in the USA and *Asparagus officinalis* in New Zealand. This prompts one to consider the possibility of hormonal effect. Later studies showed the production of IAA, cytokinin and gibberlin - like substances by AM fungi. These fungi also help in the uptake of water and protect plants against infection by soil-borne plant pathogenic fungi, bacteria and nematodes. These fungi also have synergistic interaction with beneficial soil organisms like nitrogen fixers, phosphate solubilizers and other plant growth promoting rhizomicroorganisms (PGPRs).

Anatomical and other physiological studies have brought out that mycorrhizal plants have increased rates of respiration, photosynthesis and increased amounts of sugars, amino acids, RNA etc. and larger and / or more numbers of chloroplasts, mitochondria, xylem vessels, motor cells etc. This suggests that mycorrhizal plants are much healthier with better metabolic activity. Root exudation and hence rhizosphere population (which also affects plant growth) may also be influenced by AM fungi. AM fungi increasing nodulation and nitrogen fixation by rhizobia in many grain and fodder legumes is now well established. Mycorrhizal colonization also allows introduced populations of beneficial soil organisms like *Azotobacter* and phosphate solubilizing bacteria to maintain higher numbers than around non-mycorrhizal plants, and to exert synergistic effects on plant growth. Interaction studies with plant pathogens although brought out that each pathogen-mycorrhizal fungus-plant combination is unique making it difficult to draw generalizations, however, the possibility of biologically controlling the root pathogens looks promising. All these studies bring out that AM fungi help host plant in more than one way. Whatever be the mechanism of action it is well established that AM fungal inoculation helps plant growth.

It is well known that AM fungi are not host specific. Though a particular AM fungus can infect and colonize many host plants, it has a preferred host, which exhibits maximum symbiotic response when colonized by that particular AM fungus. This led to the concept of 'host preference' in AM fungi and in turn the procedure for screening and selecting an efficient fungus for a particular host. This led to the selection of inoculant AM fungi for many crops important in agriculture, horticulture and forestry.

AM fungi are obligate symbionts. Attempts to culture AM fungi on artificial media has met with little or no success. At present time, the only method to produce these fungi is in association with the host plant root. Novel techniques to produce AM inoculum in almost sterile environment through nutrient film technique, circulatory hydroponic culture system, root organ culture and tissue culture are available. However, for large-scale field trials the only convenient method to produce large quantities of quality inoculum is by the traditional pot culture technique.

Use of Mycorrhizal Fungi for Plant Growth

AM inoculum of suitably selected strains can be used for inoculation in the nursery bed. Growers only need to incorporate inoculum in the nursery beds or seedling trays at the appropriate rate by hand. Seedlings thus raised will be colonized by the introduced fungus and then can be planted out in the field. There are several reports of increased growth and yield of food, fodder and fuel crops because of inoculation with efficient AM fungi. These studies also brought out that because of inoculation nearly 50% of phosphate fertilizer application could also be reduced. Some horticultural plants are propagated through cuttings. In such cases, rooting of cuttings is important. Enhanced rooting of cuttings through inoculation with AM fungi has been reported. AM fungi inoculated plants withstanding transplant stock have also been reported in avocado. The recent studies showed high percentage of grafting success in cashew.

Soilless media like vermiculite, perlite, potting mixes etc. are used for raising several horticultural crop plants. Such soilless media are usually fumigated or heat sterilized. Inoculation of such soilless media enhancing seedling growth and finally yield has been reported in woody ornamentals and asparagus. Inoculation of micropropagated plantlets with AM fungi after hardening also improved plantlet vigour and growth in coffee, grapevine, apple, avocado, pineapple, kiwi fruit, strawberry, raspberry, asparagus and banana.

The method of inoculation is very simple and can easily be followed by farmers. Inoculum production must be given top priority and should be undertaken by private entrepreneurs, to make the inoculum available to growers. The growers should be properly educated through demonstrations and other extension methods so that the inoculation with AM fungi can become a standard nursery practice. In the current era of organic farming, emphasis is given to reduce the use of chemicals and to promote the use of organic manures and pesticides. Thus use of AM fungi can be considered as an alternate strategy to more rational and sustainable agriculture.

Practical Applications of Mycorrhizal Fungi

Ectomycorrhizal fungi

There are numerous examples demonstrating the ability of selected, artificially inoculated mycorrhizal fungi to enhance the growth and nutritional status of tree seedlings under nursery condition and in the field following outplanting. In many parts of the world, unsuccessful attempts at afforestation and reclamation have been attributed to the absence of appropriate mycorrhizal fungi. Therefore inoculation may be particularly beneficial on reforestation sites where soil is poor and weather conditions are extreme. Further more in cases where soil-borne mycorrhizal inoculum is insufficient, for example following severe disturbance and where fungal dispersal is inadequate, out planting of seedlings preinoculated with appropriate ectomycorrhizal fungal strains should be encouraged. The soil found on degraded sites is frequently low in available nutrients, mycorrhizal fungi, and other beneficial microorganisms and inoculation with selected ectomycorrhizal fungi can significantly improve survival, growth, and establishment of young trees. Furthermore, preinoculation may enhance establishment under unfavorable environmental and soil conditions. Although the inoculation of nursery seedlings with selected ectomycorrhizal fungi is a technique generally considered for use during the production of seedlings to be outplanted in soils with low inoculum potential, inoculation has also proven beneficial for routine reforestation.

Field performance of inoculated seedlings

The fundamental means of assessing the effectiveness of inoculating the nursery seedlings are their survival and growth in the field. In Europe, forest tree seedlings inoculated with *Laccaria bicolor* S238N showed a significant increase in stem volume 8 years after outplanting. Further more in the United States, the success of *Pisolithus tinctorius* in improving seedling growth and survival after outplanting is well recognized. Successful nursery inoculation of seedlings with different mycorrhizal fungi has been widely reported in Canada. However, little is known concerning the survival and growth of outplanted mycorrhizal seedlings and only a few studies have investigated their long-term survival and success (Table 14).

Table 14: Effect of different ectomycorrhizal inoculation treatments of seedlings mean plot volume index (PVI) 5 years after outplanting.

Plant species	Inoculation treatment	PVI (cm^3) (5 yrs after outplanting)
Black spruce	Control	796
	Hebeloma longicaudum	1237
	Laccaria bicolor	1552
	Paxillus involutus	1236
	Pisolithus tinctorius	1184
	Rhizopogon vinicolor	719
	Suillus tomentosus	1435
White spruce	Control	1249
	H. longicaudum	742
	L. bicolor	465

Contd...

	P. involutus	683
	P. tinctorius	630
	R. vinicolor	719
	S. tomentosus	1111
Lodgepole pine	Control	1679
	H. longicaudum	2557
	L. bicolor	2418
	P. involutus	5622
	P. tinctorius	2831
	R. vinicolor	2759
	S. tomentosus	2543

In India, studies on taxonomy, physiology, inoculum production and application techniques of ectomycorrhizal fungi have been carried out by some workers. Not much has been done on inoculation of forest tree seedlings in the nursery with these fungi and their performance when outplanted in the field site.

Some progress has also been made, in the use of ectomycorrhizal fungi in food industry. The fruit bodies (mushrooms and truffles) of many ectomycorrhizal fungi are edible. The most highly sought fungal food is the truffle, the hypogeous fruit body of species of *Tuber*, a genus which forms ecologically obligate ectomycorrhizae on a variety of broad leaved trees. Commercial production of truffles has been pioneered in France using oak and hazel seedlings inoculated with the spores, minced fruit bodies, or fragments of established mycorrhizal roots. About 65,000 inoculated tree seedlings are now being planted annually. Pine seedlings have been inoculated with *Tricholoma matsutaki* and *T. melanosporum* with a view to be outplanted and yield edible mushrooms in the subsequent years. More research in the art of using ectomycorrhizal fungi for mushroom production is needed.

Arbuscular mycorrhizal fungi

The common mycorrhizal association in most of the plants is the AM type occurring in the majority of agricultural crops, most shrubs, most tropical tree species and some temperate tree species. Pot experiments conducted earlier in sterilized soil showed that AM inoculation can drastically improve plant growth. Later investigations indicated that even in unsterile soils plants do respond to inoculation with efficient strains of AM fungi (Plate 13). The increased growth of mycorrhizal plants is favoured in soils with low to moderate fertility. Greater soil exploration by mycorrhizal roots as a means of increasing phosphate uptake is well established. They also improve the uptake of the other elements Zn, Cu, S etc. The other beneficial effects are their role in the biological control of root pathogens, biological nitrogen fixation, hormone production and greater ability to withstand water stress. If the beneficial effects of AM fungi observed under pot culture studies could be emulated under field conditions, it would have large economic impact on agriculture.

a

b

c

Plate 13 : Response of plants to AM fungal inoculation a. Patchouli, b. Cardamom, c. Teak

Recent studies under field conditions have brought out that inoculation with efficient AM fungi increases the growth and yield of plants important in agriculture, horticulture and forestry. For this it is important to standardize the dose and time of application of AM fungal inoculum apart from the method of application in order to harness the maximum benefit from this association. Inoculation at the time of sowing / planting will result in better colonization of the roots as they emerge and hence help in better establishment of the host. The experiments conducted with micro-propagated banana, syngonium and spathiphyllum have revealed that the best time for inoculation would be just before planting out hardened plantlets.

Methods of application of AM fungi generally include hand placement, placing below the seed material or in case of pot experiments uniformly mixing the inoculum with the substrate. The importance of method of application of inoculum arises with the need to initiate colonization in the early stages of plant growth. Based on the plant parameters it was concluded that placing AM inoculum 8 cm below the surface of the soil (7 cm below the seed) at a point or as a layer is the best method of inoculum placement for seedlings raised in polybags. For seedlings raised on nursery beds placing the inoculum 2 cm below the seed was found to be the best approach.

Significant increases in plant growth and yield of several plants important in agriculture, horticulture and forestry because of AM inoculation in unsterile soils containing less or insufficient indigenous endophytes have been reported by several workers. These studies also brought out that application of phosphatic fertilizer can be reduced by nearly 50% (Table 15). Further, working with citrus rootstocks it has been observed that mycorrhizal inoculation resulted in plants ready for budding 4-5 months early compared to uninoculated plants. A similar observation was also made in mango rootstocks, wherein inoculated plants were ready for grafting two months early. Early grafting and faster establishment after mycorrhizal inoculation has been reported in cashew rootstocks as well. Again, inoculation of chrysanthemum with *Glomus mosseae* not only increased the flower yield per plant but also the vase life of cut flowers. In medicinal and aromatic plants AM fungal inoculation not only increased the crop yield but also that of the active ingredient.

Table 15: Effect of different AM fungi and added phosphorus on fruit yield of chilli

Inoculation	Yield of chilli (kg/plot) Addition of P fertilizer		
	No Addition	Half the recommended level	Recommended level (75 kg/ha)
Uninoculated	0.27	0.37	0.43
Glomus fasciculatum	0.40	0.52	-
G. albidum	0.38	0.42	-
G. macrocarpum	0.32	0.40	-
G. caledonicum	0.37	0.41	-

In recent years many high value crops are raised through tissue culture. Micropropagated plantlets cultivated on agar rooting medium are usually delicate and hence hardened before planting in the field. During the process of hardening micropropagated plantlets get acclimatized to the outside environment. During the process usually 20-40% mortality is observed. Inoculation of micropropagated plantlets with AM fungi after hardening also improved plantlet vigour and growth in coffee, grapevine, apple, avocado, pineapple, kiwifruit, strawberry, raspberry, asparagus and banana. The information available as to whether perennial plants already established in the field respond to AM fungal inoculation is very meagre. In a study it was found that ten year old mulberry plants and one and half year old papaya trees positively responded to mycorrhizal inoculation.

Forest seedlings can be raised in small seed beds or trays in suitable substrate mixed with mycorrhizal inoculum. The seedlings which become mycorrhizal can be transplanted to poly bags and then to field site or directly to the field site. This method has been tried in several experiments which produced healthy, vigorously growing seedling of *Acacia nilotica, Albizia lebbeck, Azadirachta indica, Calliandra calothyrsus, Casuarina equisetifolia, Dalbergia sissoo, D. latifolia, Leucaena leucocephala, Tamarindus indica, Tectona grandis* etc. Further it was observed that inoculated seedlings performed better when out planted in the field site including degraded forests and wastelands.

Despite the potential of AM fungi in enhancing nutrient uptake, stress tolerance, the monoculture of these organisms is not often enough to achieve maximum benefit for the plant. It is, therefore, necessary to build up a multiagent inoculum consisting of growth promoting microbial consortia. The various beneficial interactions among plant -

mycorrhizae - other growth promoting microorganisms have been well documented.

Interaction between AM Fungi and Other Beneficial Organisms

Interaction of AM fungi with symbiotic nitrogen fixers

There are several reports on the interaction between AM fungi and the legume bacterium *Rhizobium* species. These studies suggest that the interaction is synergistic; improving nodulation and AM fungal colonization, with consequential benefit to plant growth. Legumes have repeatedly been shown to require high levels of phosphate for effective nodulation and growth. It is also known that nitrogen fixation has a high phosphate requirement. While the principal effect of mycorrhiza on nodulation is undoubtedly phosphate mediated, mycorrhiza may also have other secondary effects. Such potentially limiting factors include supply of photosynthates, trace elements and plant hormones, which play an important role in nodulation and nitrogen fixation. Colonization by AM fungi has been found to increase the amount of phytoalexins in certain legume roots, which are isoflavanoid substances. Flavones are known to induce nod gene expression. These findings have paved new way for a line of research in understanding the role of AM fungi in the expression of nodulation gene in rhizobia.

Cell-free extracts of *Rhizobium* enhancing colonization of the host by AM fungi was observed by Spanish workers. It was later attributed to the presence of extra cellular polysaccharides produced by *Rhizobium,* which increased the number of entry points of AM fungi per unit length of root. It has been observed that certain nod factors stimulate mycorrhizal colonization of soybean. Field studies have shown great advantages of dual inoculation by the two symbionts. This has extra importance in the tropics because of the grain-legume programmes introduced to increase protein content of the diet and the fact that tropical soils are highly deficient in phosphorus.

The actinomycete *Frankia* is known to produce nitrogen-fixing nodules on the roots of non-legumes like *Alnus, Casuarina, Ceanothus, Myrica,* etc. The actinomycete is housed in nodules where atmospheric nitrogen is fixed and made available to the host plant. Dual inoculation with AM fungi and *Frankia* increased the total dry weight of shoots and roots, number of nodules, weight of the nodular tissues, as well as levels of N and P in *Casuarina.*

Interaction of AM fungi with asymbiotic nitrogen fixers

Nearly 25 genera of free living bacteria can fix atmospheric nitrogen. Species of *Azotobacter, Azospirillum, Beijerinckia, Clostridium* and *Derxia* are well known among these. Bagyaraj and Menge (1978) studied the interaction between *Azotobacter chroococcum* and the AM fungus *Glomus fasciculatum* in tomato and found a synergistic effect on plant growth. Mycorrhizal infection increased the *A. chroococcum* population in the rhizosphere, which was maintained at a high level for a longer period and *A. chroococcum* enhanced colonization and spore production by the mycorrhizal fungus. Similar interactions have also been observed between *A. paspali* and AM fungi in paspalum and between *A. chroococcum* and *G. fasciculatum* in tall fescue. Synergistic interaction between AM fungi and associative nitrogen fixing bacteria *Azospirillum* spp. and *Acetobacter diazotrophicus* have been reported.

A tripartite interaction study between the free-living nitrogen fixing bacterium *Beijerinckia mobiles,* phosphate solubilizing fungus *Aspergillus niger* and the mycorrhizal fungus *G. fasciculatum* was conducted. A synergistic beneficial effect on the growth of onion with all 3 organisms was observed. These were attributed to hormone production rather than, or in addition to nitrogen fixation. The hormones produced by these bacteria could exert some synergistic effect on plant growth or mycorrhizal efficiency. The studies conducted so far reveal a definite positive interaction between free-living nitrogen fixing bacteria and AM fungi in the rhizosphere with consequential improvement in plant growth.

Interaction of AM fungi with phosphate solubilizers

Many soil bacteria termed as phosphobacteria solubilize unavailable forms of phosphorus and have been used as bacterial fertilizers on crop plants. Interaction between AM fungi and phosphate solubilizing microorganisms and their effect on plant growth have been studied by several workers. Phosphate solubilizing bacteria (*Agrobacterium* sp. and *Pseudomonas* sp.) inoculated onto the seeds and / or seedlings maintained higher populations for longer duration in the rhizosphere of mycorrhizal than non-mycorrhizal roots of lavender and maize plants. Dual inoculation also resulted in increased plant dry matter and phosphorus uptake in soils. The phosphate solubilizing bacteria also produced plant growth hormones which enhanced plant growth.

The effect of *Glomus fasciculatum* and non-phytohormone producing strain of P solubilizer *Bacillus circulans* on phosphate solubilization, growth of finger millet and phosphorus uptake from ^{32}P- labeled tricalcium phosphate and superphosphate was studied. The results clearly revealed that AM fungi did not solubilize unavailable form of phosphorus but still enhanced the P uptake, which was attributed to better exploration of soil. Studies conducted with neem and *Pennisetum* grass further confirmed a synergistic interaction between AM fungi and phosphate solubilizing bacteria.

Interactions between AM fungi and mycorrhiza helper organisms

It has been found that some microorganisms present in soil, can modify the establishment of the mycorrhizal symbiosis. The organisms associated with the rhizosphere of mycorrhiza which, stimulate the growth of the fungus and mycorrhizal colonization are called mycorrhiza helper organisms (MHO). Therefore, MHO selectively helps AM fungal species to form the symbiotic association by increasing its competitive ability against other potential symbionts. It was suggested that it could be due to either the production of enzymes or growth promoting substances which increase the cell wall plasticity and thereby increase fungal colonization.

Several workers have reported that co-inoculation of AM fungi with MHO enhanced mycorrhizal colonization and in turn plant growth and yield. In high-tech agriculture some crops are raised in disinfected soil / nursery. Disinfection kills several groups of organisms. In such soils, mycorrhizal inoculation is recommended. If MHO are also killed then colonization by AM fungi will not be proper. It is suggested that milder disinfection treatments which selectively kill indigenous AM fungi and root pathogens but conserving MHO should be developed. It is probable that, in the near future, commercial mycorrhizal inoculum will contain MHO, improving the efficiency of inoculation in a wider range of conditions and reducing the quantity of inoculum needed.

Recent studies have shown that inoculation with microbial consortia consisting of an efficient AM fungus together with a nitrogen fixer, P solubilizer and PGPR carefully screened and selected for a particular crop plant or forestry species is more beneficial than AM fungus alone in improving the growth, biomass and yield (Plate 14).

Plate 14 : Response of a. *Piper longum*, b. *Acacia auriculiformis* to inoculation U – Uninoculated, M – AM fungus, MC – Microbial consortia

The fundamental means of assessing the effectiveness of inoculating the nursery seedlings are their survival and growth in the field. Such studies have been done with ectomycorrhizal fungi mostly in the USA and Canada. The success of ectomycorrhizal fungal inoculation in improving seedling growth and survival after out planting is well recognized. However such studies with AM fungi alone or as microbial consortia are megre. In a recent study forest tree seedlings inoculated in the nursery with microbial consortia consisting of selected AM fungi and PGPRs were planted in degraded forest. Inoculated plants survived and grew much better compared to uninoculated plants 52 months after planting (Plate 15; Table 16).

Plate 15 : Growth of a. *Tectona grandis* b. *Acacia auriculiformis* planted in degraded forest 52 months after planting
U – Uninoculated; I – Inoculated with microbial consortia

Table 16: Response of *Acacia auriculiformis* and *Tectona grandis* to microbial inoculation in degraded forests at three locations 52 months after planting

Location	Biovolume index [Plant height (cm) x Stem girth (mm)]	
	Acacia auriculiformis	*Tectona grandis*
Mandya		
Uninoculated	10,390	746
Inoculated	20,479	3,556
Srirangapatna		
Uninoculated	11,661	944
Inoculated	17,889	1,818
Pandavapura		
Uninoculated	7,081	216
Inoculated	11,433	587

References

Bagyaraj, D.J. and Menge, J.A. 1978. Interaction between a VA mycorrhiza and *Azotobacter* and their effects on the rhizosphere microflora and plant growth. *New Phytologist* 80:567-573.

Bagyaraj, D.J., Krishnaraj, P.U. and Khanuja, S.P.S. 2000. Mineral phosphate solubilization: Agronomic implications, mechanism and molecular genetics. *Proceedings of Indian National Science Academy. (PINSA)*. B 66: 69-82.

Douds Jr.D.D., Nagahashi, G., Pfeffer, P.E., Kayser, W.M. and Reider, C. 2005. On-farm production and utilization of arbuscular mycorrhizal fungus inoculum. *Canadian Journal of Plant Sciences,* 85:15-21.

Duponnois, R., Colombet, A., Hein, V. and Thioulouse, J. 2005. The mycorrhizal fungus *Glomus intraradices* and rock phosphate amendment influence plant growth and microbial activity in the rhizosphere of *Acacia holosericea. Soil Biology and Biochemistry,* 37:1460-1468.

Foth, H.D. and Ellis, B.G. 1997. *Soil Fertility* (2nd edn.). Lewis Publishers, CRC Press, Boco Raton, Florida, USA.

Gaind, S., Mathur, R.S. and Tilak, K.V.B.R.. 2001. Phosphate solubilizing microorganisms. In: *Recent Advances in Biofertilizer Technology* (eds. A.K.Yadav, M.R. Motsara and S. Ray Chaudhuri), Society for promotion & Utilization of Resources and Technology, New Delhi, pp. 190-206.

Gaur, A.C. 1985. Phosphate solubilizing microorganisms and their role in plant growth and crop yield. In: *Proceedings of Soil Biology Symposium,* HAU, Hissar, Haryana, India. pp. 125-138.

Gaur, A.C. 1990. *Phosphate Solubilizing Microorganisms as Biofertilizers.* Omega Scientific Publishers, New Delhi, India.

Guar, A. and Adholeya, A. 2004. Prospects of arbuscular mycorrhizal fungi in phytoremediation of heavy metal contaminated soils. *Current Science,* 86:528-534.

Gyaneshwar, P., Kumar, G.N., Parekh, L.J. and Poole, P.S. 2002. Role of soil microorganisms in improving P nutrition of plants. *Plant and Soil,* 245 :83-93.

Igual, J.M., Valverde, A., Cervantes, E. and Velazoez, E. 2001. Phosphate-solubilizing bacteria as inoculants for agriculture: use of updated molecular techniques in their study. *Agronomie,* 21:561-568.

Illumer, P. and Schinner, F. 1995. Solubilization of inorganic calcium and phosphorus: Solubilization mechanisms. *Soil Biology and Biochemistry,* 27:257-261.

Jeffries, P. Gianinazzi, S., Perotto, S., Turnau, K. and Barea, J.M. 2003. The contribution of arbuscular mycorrhizal fungi in sustainable maintenance of plant health and soil fertility. *Biology and Fertility of Soils,* 37:1-16.

Khade, S.W. and Rodrigues, B.F. 2003. Occurrence of arbuscular mycorrhizal fungi in tree species from Western Ghats of Goa, India. *Journal of Tropical Forest Science,* 15:320-331.

Khasa, D., Piche, Y. and Coughlan, A.P. 2009. *Advances in Mycorrhizal Science and Technology.* NRC Research Press, Ottawa, Canada.

Koide, R.T. and Mosse, B. 2004. A history of research on arbuscular mycorrhiza. *Mycorrhiza,* 14:145-163.

Li, X.L. and Christie, P. 2001. Changes in soil solution Zn and pH and uptake of Zn by arbuscular mycorrhizal red clover in Zn-contaminated soil. *Chemosphere,* 42:201-207.

Mehrotra, V.S. 2005. *Mycorrhiza - Role and Applications.* Allied Publishers Pvt. Ltd., New Delhi, India.

Mehta, S. and Nautiyal, C.S. 2001. An efficient method for qualitative screening of phosphate-solubilizing bacteria. *Current Microbiology,* 43:51-56.

Mikanova, O. and Novakova, J. 2002. Evaluation of P solubilizing activity of soil microorganisms and its sensitivity to soluble phosphate. *Rostlinna Vyroba,* 48:397-400.

Mukerji, K.G., Chamola, B.P. and Singh, J. 1999. *Mycorrhizal Biology.* Kluwer Academic/Plenum Pub., New York, USA.

Muthu Kumar, T. and Bagyaraj, D.J. 2010. Use of mycorrhizal fungi in phytoremediation of heavy metal contaminated soils. *Proceedings of the National Academy of Sciences, India.* Section B.80: 103-121.

Paul, E.A and Clark, F.E. 1989. *Soil Microbiology and Biochemistry.* Academic Press, New York, USA.

Pikovskaya, R.I. 1948. Mobilization of phosphorus in soil in connection with the vital activity of some microbial species. *Microbiologiya,* 17:362-370.

Podila, G.K. and Varma, A. 2005. *Basic Research and Applications of Mycorrhizae.* I.K.International Pvt. Ltd., New Delhi, India.

Powell, C.L. and Bagyaraj, D.J. 1984. *VA Mycorrhiza.* CRC Press, Inc. Boca Raton, Florida, USA.

Read, D.J. 1991. Mycorrhizas in ecosystem. *Experientia,* 47:376-391.

Rillig, M.C., Wright, S.F. and Eviner, V.T. 2002. The role of arbuscular mycorrhizal fungi and glomalin in soil aggregation: comparing effects of five plant species. *Plant and Soil,* 238:325-333.

Sanders, F.E., Mosse, B. and Tinker, P.B. 1975. *Endomomycorrhizas*. Academic Press, London.

Schenck, N.C. and Perez, Y. 1990. *Manual for Idendification of VA Mycorrhizal Fungi*, 3rd edition, Synergistic Pub., Gainesville, USA.

Sharma, A.K. and Johri, B.N. 2002. *Arbuscular Mycorrhizae- Interactions in Plants, Rhizosphere and Soils*. Oxford & IBH Publishing Co. Pvt. Ltd., New Delhi, India.

Singh, R. and Adholeya, A. 2002. Biodiversity of arbuscular mycorrhizal fungi (AMF) in different agroclimatic regions of India. IMC 7th Conference on Mycological Advances, 11-17August 2002, Norway.

Whitelaw, M. 2000. Growth promotion of plants inoculated with phosphate-solubilizing fungi. *Advances in Agronomy*, 69:99-151.

Chapter-5

Plant Growth Promoting Rhizomicroorganisms

The rhizosphere is the narrow zone of soil surrounding the root that is under the immediate influence of the root system. This zone is rich in nutrients when compared with the bulk soil, due to the accumulation of a variety of organic compounds released from roots by exudation, secretion and deposition. Organic compounds released by plant roots include amino acids, fatty acids, nucleotides, organic acids, phenolics, plant growth regulators, putrescine, sterols, sugars and vitamins. Because these organic compounds can be used as carbon and energy source by microorganisms, microbial growth and activity is particularly intense in the rhizosphere. This is reflected by the number of bacteria that are found around the roots of plants and that is generally 10 to 100 times higher than in the bulk soil. Plant associated bacteria that are able to colonise the roots are called rhizobacteria and can be classified into beneficial, deleterious, and neutral groups on the basis of their effects on plant growth. Beneficial rhizobacteria that stimulate plant growth are usually referred to as Plant Growth Promoting Rhizobacteria (PGPR). PGPR strains are often found among the fluorescent pseudomonads, rhizobia, *Acetobacter, Actinoplanes, Agrobacterium, Alcaligenes, Arthrobacter, Azospirillum, Azotobacter, Bacillus, Cellulomonas, Enterobacter, Flavobacterium, Pasteuria, Serratia, Pantoea, Paenibacillus* etc. (Plate 16).

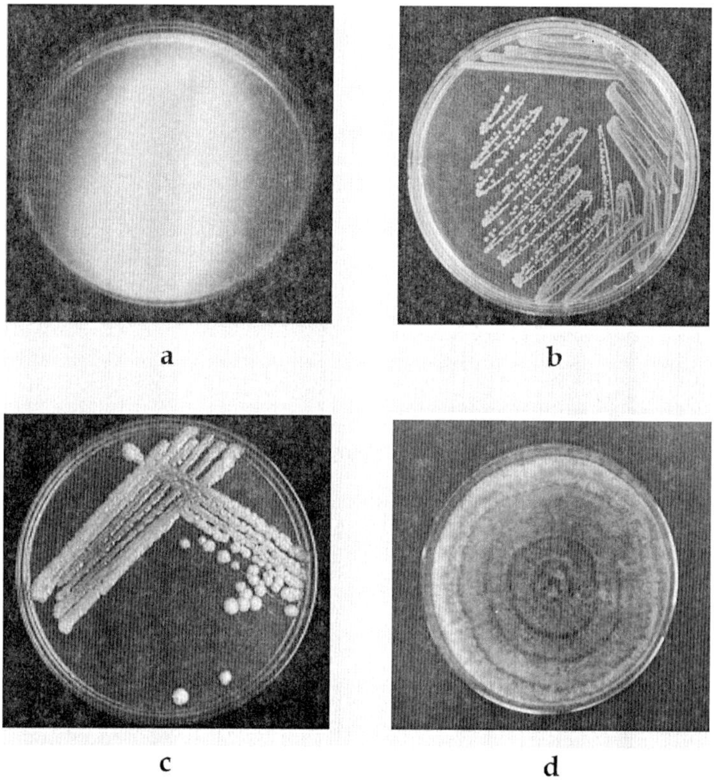

Plate 16 : Plates showing growth of PGPRs a. *Pseudomonas fluorescens,* b. *Pantoea agglomerans* c. *Paenibacillus polymyxa* d. *Trichoderma harzianum*

Later it was observed that certain filamentous fungi like *Trichoderma harzianum, Piriformospora indica* and also yeasts like *Saccharomyces cerevesiae, Rhodotorula mucilagina* also enhance plant growth (Boby *et al.*, 2008). Hence the abbreviation PGPR is now expanded as plant growth promoting rhizomicroorganisms. This change was proposed and accepted during the 1st Asian congress on PGPRs held at Hyderabad in 2009.

The first reports of PGPR were confined to root crops such as radish, potato and sugarbeet. Subsequent field studies have revealed that species of *Pseudomonas* and similar bacteria increased the yield of potato (up to 33%), sugarbeet (up to 15%) and radish (up to 144% of root weight) when seed or seed pieces were treated with selected bacterial strains before planting. Subsequent reports indicate that PGPR can exert positive effects on diverse hosts, including barley, bean, canola (mustard), cotton, peanut, lentil, rice, soybean, vegetables and woody species including apple and citrus. These growth promoting rhizobacteria have been studied in relation to a wide range of crop plants like cereals, pulses, oil seeds, fibre crops, vegetable and horticultural crops, plantation crops, etc.

Mechanisms of Plant Growth Promotion

PGPR can affect plant growth in two different ways, directly or indirectly. The direct promotion of plant growth by PGPR for the most part entails either providing the plant with a compound that is synthesised by the bacterium or facilitating the uptake of certain nutrients from the environment. The indirect promotion of the plant growth occurs when PGPR lessens or prevent the deleterious effect of one or more phytopathogenic organisms. The mechanisms by which PGPR promote plant growth have been reviewed by several workers. It has been suggested that production of plant hormones, enhanced nutrient uptake, and suppression of phytopathogenic microorganisms *via* siderophores, HCN, β-1,3- gluconases, chitinases, secretion of ammonia, production of antibiotics and volatile substances are involved.

Mechanisms of Direct Growth Promotion

Production of plant growth regulators

Phytohormones, also called plant growth regulators (PGRs), are well known for their regulatory role in plant growth and development. PGRs are organic substances that influence physiological processes of plants at extremely low concentrations. Because the concentration of hormonal signals is critical to the regulation of various physiological processes in plants, local changes of phytohormone levels can lead to characteristic changes in plant growth and development.

The phytohormones produced by PGPRs are mostly indole acetic acid, cytokinins, gibberellins and a few others like abscisic acid (ABA), zeatin riboside etc, which often lead to morphological changes in the plants. The bacterial strains regulate plant growth by mimicking synthesis of plant hormones including indole-3-acetic acid, cytokinins, or gibberellins while others increase mineral and nitrogen availability in the soil. Plants differ in the levels and ratio of the hormones required to maintain normal growth and development therefore, it might be expected that different stages of development respond differently to invasion by hormone producing bacteria.

IAA producers

The first evidence for the role of IAA in the observed plant growth promotion was obtained by attempts to mimic the effect of the bacterium on root growth by the direct application of IAA onto the roots. It could be shown with wheat that inoculation with *Azospirillum brasilense* and the application of pure IAA to the roots both increased root length, number of lateral roots and number of root hairs. Similar experiments with *A. brasilense* and *Paenibacillus polymyxa* provided strong evidence for IAA production by these diazotrophs and the responsibility of this hormone for the observed effects on plants. *Pseudomonas savastanoi* has been reported to produce large amounts of IAA when grown in presence of tryptophan. This bacterium causes tumor like outgrowths in olives. The capacity to induce tumors is correlated with capacity to produce auxins which indicate that auxin is major determinant in virulence and also suggest that microbially released plant growth promoting substances in the rhizosphere affect plant growth.

Azospirillum brasilense was found to produce extremely high amounts of IAA. However, the amounts of IAA produced depended on the species and strains as well as on the conditions of cultivation. In wheat, the inoculation with IAA - producing *A. brasilense* and application of pure IAA to the roots was found to increase root length, number of lateral roots and numbers of root hairs. The mutant strains of *A. brasilense* that synthesised very low levels of IAA, no longer promoted the formation of lateral roots of wheat seedlings when compared with the wild-type strain, indicating the role of bacterially synthesised IAA in promotion of plant growth.

In addition to IAA, bacteria such as *P. polymyxa* and azospirilla also release other compounds in the rhizosphere that could indirectly contribute to plant growth promotion like indole - 3 - butyric acid (IBA), trytophan, tryptophol (TOL) or indole-3-ethanol. IBA is a compound that is widely used in agriculture as a commercial promoter of root initiation in cuttings. Tryptophan may enhance endogenous IAA synthesis in the plant root. Direct uptake of applied tryptophan by plant roots followed by conversion into IAA within their tissues has already been proposed. In addition to the synthesis of IAA, plant growth promotion by *P. polymyxa* strains was proposed to be due to its production of TOL. TOL is formed after the reduction of indole - 3 - acetaldehyde, a side reaction of the IAA pathway. It is easily taken up by plants, serves as an IAA storage form, and can be converted into the active phytohormone, IAA, by plant TOL - oxidase and O_2.

Cytokinin producers

The isolation and quantification of cytokinins in non-pathogenic soil bacteria has received little attention. Because cytokinins are a diverse group of labile compounds that are usually present in small amounts in biological samples, they have often been difficult to identify and quantify. Usually, cytokinin-like activity is reported using bioassays. The effects of exogenously applied cytokinins on plants are numerous, the most notable of which is enhanced cell division, but also root development and root hair formation are reported. Plants and plant-associated microorganisms have been found to contain over 30 growth-promoting compounds of the cytokinin group. One study indicated that as many as 90% of the microorganisms found in the rhizosphere are capable of releasing cytokinins when cultured *in vitro.* Using bioassays, radio-immunoassays, or HPLC-UV spectrometry, cytokinin production was shown in *Azotobacter* spp., *Azospirillum* spp., *Rhizobium* spp. and *P. polymyxa.*

The influence of the cytokinin precursors adenine (ADE) and isopentyl alcohol (IA) in the cytokinin producing bacterium *Azotobacter chroococcum* and their effect on the morphology and growth of radish and maize under *in vitro,* green house and field conditions has been studied. The combination of ADE, IA plus the bacterium enhanced the growth of the plants to a much greater degree than when only the precursors or *A. chroococcum* were applied. The improvement in plant growth was attributed primarily to the increase in cytokinin production by *A. chroococcum* in the rhizosphere. As cytokinins move from roots to shoots, root exposure to cytokinins could affect plant growth and development. Increases in yield and N, P and K content of grains obtained after exogenous application of cytokinins in field trials with rice support the hypothesis that bacterially supplied cytokinins to the soil can improve the growth and yield of treated plants.

Gibberellin producers

Over 89 gibberllins (GAs) are known to date and are numbered GA_1 through GA_{89} in approximate order of their discovery. The most widely recognised gibberellin is GA_3 (gibberellic acid), the most active GA in plants is GA_1, which is primarily responsible for stem elongation. Using a bioassay, HPLC or GC-MS, GA production has been demonstrated in *Azotobacter* spp., *Paenibacillus polymyxa, Rhizobium leguminosarum* bv. phaseoli, *Azospirillum brasilense, Azotobacter lipoferum, Acetobacter diazotrophicus* and *Herbaspirillum seropedicae, Bacillus pumilus* and *B. licheniformis*. Most likely, the GAs formed do not play a significant physiological role in the producing organism and, consequently can, be regarded as secondary metabolites. However, several observations suggest that these GA- producing microorganisms might induce or promote growth in the host plants through the action of released GAs. One of the easiest studies in this respect reports on similar effects obtained by inoculation of tomatoes with *Azotobacter chroococcum* or gibberellin application. GA_3 had similar effects as *A. lipoferum* inoculation on promotion of root growth in 48 hour old maize seedlings, especially in increasing hair density in areas physiologically active for nutrient uptake and water absorption. *A. brasilense, Bacillus pumilus* and *B. licheniformis* enhancing growth of maize and rice because of production of GAs has also been reported.

The ability of *Azospirillum* spp. to alleviate the effects of water deficits in cereal seedlings under salt and osmotic stresses can also be attributed at least partly to bacterial GA production. This increased the chlorophyll, K, Ca, soluble sugars, protein content and improvement of the active hairy root zone facilitating water absorption because of gibberellin produced by the bacterium.

When studying role of phytohormone production in the mechanism of plant growth promotion by microorganisms, some caution is required concerning the interpretation of the results. Since plants as well as many microorganisms can synthesise phytohormones, one has to take into account that microorganisms may also be able to affect plant endogenous hormone levels, irrespective of their own hormone production. It is hard to distinguish between the hormones synthesised by the plant in response to PGPR stimulation and the hormones synthesised by the PGPR itself. Therefore, all possibilities, the plant synthesising more phytohormones upon presence of the bacterium, the bacteria producing and exporting more phytohormones under the influence of plant roots, or both, remain open.

In addition, phytohormones do not act alone but rather interact with one another in a variety of complex ways. As an example, the balance between auxin and cytokinin levels controls cellular differentiation and organogenesis in tissue and organ culture, ranging from shoot proliferation to root formation as the ratio auxin/cytokinin increases. Additionally, auxins and cytokinins stimulate ethylene production synergistically. It is clear that the effect of bacterial phytohormone synthesis on plant growth is quite complex and should be seen in view of the interaction with other phytohormones.

Enzymes Modulating Plant Growth and Development

Ethylene

A rather new and unsuspected mechanism of plant growth promotion involves the plant hormone ethylene. Ethylene is a potent plant growth regulator that affects many aspects of plant growth, development and senescence. In addition to its recognition as a "ripening hormone", ethylene promotes adventitious root and root hair formation, stimulates germination, and breaks the dormancy of the seeds. However, if the ethylene concentration remains high after germination, root elongation (as well as symbiotic N_2 fixation in leguminous plants) is inhibited.

It has been proposed that many plant growth promoting bacteria may promote plant growth by lowering the levels of ethylene in plants. This is attributed to the activity of the enzyme 1-aminocyclopropane-1-carboxylate (ACC) deaminase, which hydrolyzes ACC, the immediate biosynthetic precursor of ethylene in plants. The products of this hydrolysis, ammonia and α-ketobutyrate, can be used by the bacterium as a source of nitrogen and carbon for growth. In this way the bacterium acts as a sink for ACC and as such is lowering the ethylene level in plants, preventing some of the potentially deleterious consequences of high ethylene concentrations.

An example of such an ACC deaminase containing bacterium is the PGPR *Pseudomonas putida* GR12-2 that stimulates root growth of a number of different plants (canola, lettuce, tomato) under gnotobiotic conditions. *Pseudomonas putida* GR12-2 was chemically mutagenized and three independent mutants that lacked ACC deaminase activity were selected. Unlike the wild type, none of these selected mutants was able to promote the growth of canola seedling roots under gnotobiotic conditions. It was concluded that the ACC deaminase may play a role in the mechanism that this bacterium uses to stimulate canola root elongation. The bacteria *Azospirillum* spp., *Herbaspirillum* spp., *Azoarcus* spp., *Pseudomonas putida, Enterobacter cloacae* have been reported to produce the enzyme ACC deaminase. However, the agronomic use of PGPR that promote plant growth by mechanisms that include hydrolysing ACC may be limited largely to dicots. Monocots are less sensitive to ethylene and therefore are less responsive to these bacteria.

Increased uptake of nutrients

Bacteria capable of fixing atmospheric nitrogen thus providing nitrogen to the plants also qualify, as PGPRs. Similarly microorganisms solubilizing unavailable P into a form available for plant growth are also PGPRs. Since microorganisms modulating nitrogen and phosphorus nutrition of plants is discussed else where in detail they are not discussed here.

Potassium mobilising bacteria

Potassium is a major plant nutrient and it is essential for several growth processes. Potassium is removed in large amounts by crop plants. Potassium in soil can be readily available, slowly available or difficultly available for plant growth. The difficultly available portion or fixed potassium ranges from 90 to 97% of the total soil potassium. The slowly available potassium constitutes about 2-9% of the total soil potassium, thus the readily available form is meagre. A bacterium has been isolated from banana rhizosphere which is capable of mobilising the difficultly/slowly available form to a form usable by plants. The bacterium has been identified as *Frateuria aurantia*. The bacterium belongs to the family Pseudomonadaceae. It is Gram-negative, rod shaped motile bacterium. It grows well on glucose yeast extract agar medium. The optimum temperature for growth is 25-30°C and grows well at pH 7. Several studies have been carried out under pot culture and field conditions to know the effect of *F. aurantia* on plant growth. The results showed 10-12% increased yield and also up to 50 % saving of potassium fertilizer.

Sulphur oxidising bacteria

Oxidation of sulphur to sulphate in soil is a slow process. Sulphur oxidising bacteria bring about this oxidation. Canola productivity decreases particularly in sulphur deficient soils. Significant increase in the performance and production of canola by use of sulphur oxidising bacterial inoculant *Delftia acidovorans* (BioBoost) isolated from Canadian soil, which also possesses PGPR abilities has been reported. Seed analysis showed that BioBoost inoculants helped in canola S-uptake but did not change the seed quality traits like oil, protein, oleic acid, linolenic acid and glucosinolate content of canola seed.

Vitamin producers

Vitamins are sometimes added to the list of compounds involved in direct plant growth promotion that PGPR can produce. However, the possibility that plant growth can be improved by inoculation with vitamin-producing bacteria has received little attention. It is generally accepted that green plants under optimal growing conditions synthesize sufficient quantities of vitamins to meet their own needs, but stress, induced by, for example, drought, unfavourable temperature, or mineral deficiency, might lead to vitamin deficiency, a deficiency that could in part be the cause of the observed yield reduction. Vitamins can reverse the negative effects of mineral deficiencies, but also under normal conditions growth stimulation and yield increases were frequently observed after vitamin application.

Azotobacter, Azospirillum and *Rhizobium* strains were found to produce some or all of the water-soluble B-group vitamins niacin, pantothenic acid, thiamine, riboflavine, and biotin in defined culture media. Nevertheless, their possible role in plant growth promotion has never been studied. There is evidence that exogenous B vitamins can be absorbed by roots, producing favourable effects on root development (thiamin is considered to be a growth factor for plant roots), shoot length, dry matter production, and nutrient uptake.

Azotobacter vinelandii, when grown in conditions similar to its natural habitat, produced large amounts of vitamins in the presence of combined N and when a high concentration of glucose was added as a C source. Based on these data, it can be assumed that vitamins produced by *Azospirillum* and *Azotobacter* spp. may stimulate plant growth in the field under certain C:N ratios. Comparative studies with appropriate mutants are needed to verify this hypothesis.

Mechanisms of Indirect Growth Promotion

Production of siderophore

Availability of iron, one of the most important nutrient required for the growth of almost all living organisms, is extremely limiting in the rhizosphere. To survive in such an environment organisms were found to secrete Fe-binding ligands called siderophores having high affinity to sequester Fe from the microenvironment. One way that PGPR can prevent the proliferation of phytopathogens, and thereby facilitate plant growth, is through the production and secretion of siderophore with a very high affinity for Fe. The secreted siderophore molecules bind most of the Fe^{3+} that is available in the rhizosphere, and as a result effectively prevent any pathogens in its immediate vicinity from proliferating because of a lack of Fe. Kloepper *et al.* (1980) were the first to demonstrate the importance of siderophore production as a mechanism for plant growth promotion and biological control of pathogens. They demonstrated the siderophore mediated antagonism of *Erwinia carotovora* by plant growth promoting fluorescent *Pseudomonas* strains A1, BK1, TL3, B1 and B10 and yield increases in case of potato, sugarbeet and radish. Pyoverdin and azotobactin are examples of siderophores produced by *Pseudomonas fluorescens* and *Azotobacter vinelandii* respectively. Certain siderophores as pyoverdin and salicylate may act as elicitors for inducing systemic resistance.

Production of antibiotics

One of the most effective mechanisms that a PGPR can employ to prevent proliferation of phytopathogens is the synthesis of antibiotics. Most antibiotics belong to the class of nitrogen containing heterocycles such as phenazines and pyrrolnitrin-type antibiotics. The purified antibiotics pyoluteorin and pyrrolnitrin, obtained from *P. fluorescens* Pf-5, could suppress the damping off of cotton by *Pythium ultimum or Rhizoctonia solani* like that of producing strains.

Fluorescent pseudomonads have revolutionized the field of biological control of soil-borne plant pathogenic fungi. During the last three decades, they have emerged as the largest and potentially most promising group of plant growth promoting rhizobacteria involved in the biocontrol of plant diseases. Fluorescent pseudomonads have received the most attention for several compelling reasons as they readily colonise roots in nature. The simple nutritional requirements and the ability to use many carbon sources that exude from roots and to compete with indigenous microflora, may explain their ability to colonise the rhizosphere. Additionally, pseudomonads are amenable to genetic manipulation. These characteristics make them useful vehicle for the delivery of antimicrobial and insecticidal compounds and plant hormones to the rhizosphere.

Production of lytic enzymes

A number of alternate mechanisms have also been observed for increased plant growth and resistance to disease in plants inoculated with biocontrol bacteria. The production of potent extracellular lytic enzymes such as chitinases and laminarinase, capable of destroying fungal cell walls of *Fusarium* sp. and other fungal pathogens, by some isolates of rhizospheric bacteria as *Pseudomonas stutzeri, Serratia marcescens* etc. (cultures or cell free extracts) is a possible mechanism of disease control.

The mechanism of action by plant growth promoting actinobacteria *i.e.* parasitism of pathogenic fungal spores by rhizospheric actinomycetes due to production of extracellular enzymes has also been reported. A later work also established the production of α-1-3, α-1-4 and α-1-6 glucanases from a number of actinomycete isolates that were able to suppress raspberry root rot causing fungal pathogen *Phytophthora fragariae* var. *rubi.*

Production of HCN and ammonia

Hydrocyanic acid (HCN) is produced by many rhizobacteria and is postulated to play a role in biological control of pathogens. Cyanide produced by *Pseudomonas fluorescens* strain CHAO directly antagonises *Thielaviopsis basicola* causing black root rot of tobacco. The volatile compound such as ammonia produced by *Enterobacter cloacae* was involved in the suppression of *Pythium ultimum* induced damping off of cotton.

Competition for nutrients and niches

Competition of nutrients supplied by root and seed exudates occurs between beneficial rhizobacteria and pathogens on the root and is responsible to a small degree for the observed biocontrol by introduced bacteria. There is a correlation between the ability of various bacterial strains to compete for nutrients with germinating oospores of pathogenic fungi and their efficacy as biocontrol agents against pathogen attack and appearance of disease, which suggest that large populations of bacteria established on planting material and roots become a partial sink for nutrients in the rhizosphere, thus reducing the amount of carbon and nitrogen available to stimulate germination of spores of fungal pathogens or for subsequent colonisation of the root. Fluorescent pseudomonads are especially suited for rapid uptake or scavenging of nutrients, since they are nutritionally versatile and grow rapidly in the rhizosphere. Niche exclusion is potentially an important mechanism of antagonism of deleterious rhizobacteria by PGPR as the products secreted by PGPR exhibit exclusion of harmful rhizomicrobes *via* antagonistic interactions.

Induction of systemic resistance

Interaction of some bacteria with the plant roots can result in plants resistant to some pathogenic bacteria, fungi and viruses. This phenomenon is called induced systemic resistance (ISR). ISR shares many properties with innate immunity in humans. ISR differs from SAR (systemic acquired resistance). ISR was discovered by the findings that resistance could be induced by the rhizobacterium *Pseudomonas* sp. strain WCS417r against *Fusarium* wilt of carnation and by selected rhizobacteria against the fungus *Colletotrichum orbiculare* in cucumber. ISR is dependent on jasmonic acid and ethylene signalling in the plant. Many individual bacterial components induce ISR, such as LPS, flagella, salicylic acid and siderophores. More recently, cyclic lipopeptides, the antifungal factor 2,4-diacetyl phloroglucinol (Phl), the signal molecule N-acyl homoserine lactone (AHL), and volatile blends produced by *B. subtilis* GB03 and, to a lesser extent, the individual volatiles acetoin and 2,3-butanediol have been added to the list. In contrast to many biocontrol mechanisms, extensive colonization of the root system is not required for ISR, as shown by the ISR of *P. fluorescens* WCS365 using root colonization mutants.

Use of PGPR as Biofertilizer

The nitrogen fixing and phosphate solubilizing bacteria have been discussed separately in this book. The other PGPR organisms under most scrutiny for potential use in agriculture are bacteria belonging to the genera *Pseudomonas* and *Bacillus*. Inoculation of three PGPR isolates each belonging to *Proteus vulgaris, Klebsiella planticola* and *Bacillus subtilis* markedly increased the seed yield of sunflower, maize and cotton over uninoculated plants grown in field.

Treatment of wheat seeds with fluorescent pseudomonads (antagonistic to *Gaeumannomyces graminis*) resulted in yield increases of 27% in field trials. Researches indicate that combinations of PGPR strains (two or more) which have diverse mode of plant growth promotion or antagonism against soil-borne pathogens are more effective than single strain inoculum. Field studies also indicated that PGPR strains in conjunction with *Rhizobium* or *Azospirillum* increased the grain yield of chickpea (Table 17) and wheat, respectively. The effect was quite conspicuous at lower levels of nitrogen (40 kg N / ha) in case of wheat.

Table 17: Field response of chickpea to inoculation with PGPRs

Treatment	Grain yield (kg/ha)
Control	2,145
Bacillus subtilis (Bs)	2,171
Klebsiella planticola (Kp)	2,196
Proteus vulgaris (Pv)	2,248
Rhizobium (R)	2,313
R +Bs	2,326
R +Kp	2,416
R + Pv	2,481
CD (P = 0.05)	247

References

Balandreau, J. and Knowles, R. 1978. The rhizosphere. In: *Interactions Between Non-pathogenic Soil Microorganisms and Plants* (eds. Y.R.Dommergues and S.V. Kurpa), Elsevier, Amsterdam, Holland. pp. 243-268.

Barber, D.A. 1976. Nutrient uptake. In: *Interactions between Non-pathogenic Soil Microorganisms and Plants* (eds. Y.R.Dommergues and S.V. Kurpa), Elsevier, Amsterdam, Holland. pp.131-162.

Bassler, B.L. 1999. How bacteria talk to each other: regulation of gene expression by quorum sensing. *Current Opinions in Microbiology,* 2:582-587.

Boby, V.U., Balakrishna, A.N. and Bagyaraj, D.J. 2008. Interaction between AM fungus *Glomus mosseae* and soil yeasts on growth and nutrition of cowpea. *Microbiological Research,* 163:693-700.

Cazorla, F.M., Duckett, S.B., Bergstrom, E.T., Noreem, S., Odijk, R., *et al.* 2006. Biocontrol of avocado *Dematophora* root rot by the antagonistic *Pseudomonas fluorescens* PCL1606 correlates with the production of 2-hexyl 5-propyl resorcinol. *Molecular Plant Microbe Interactions,* 19:418-28.

Chin-A-Woeng, T.F.C., Bloemberg, G.V., Mulders, I.H.M., Dekkers, L.C. and Lugtenberg, B.J.J. 2000. Root colonization is essential for biocontrol of tomato foot and root rot by the Phenazine-1-carboxamide-producing bacterium *Pseudomonas chlororaphis* PCL1391. *Molecular Plant Microbe Interactions,* 13:1340-1345.

Choi, O., Kim, J., Kim, J.G., Jeong, Y., Moon, J.S., Park, C.S., and Hwang, I. 2008. Pyrroloquinoline quinine is a plant growth promotion factor produced by *Pseudomonas fluorescens* B16. *Plant Physiology,* 146:657-68.

Compant, S., Duffy, B., Nowak, J., Clement, C. and Barka, E.A. 2005. Use of plant growth promoting bacteria for biocontrol of plant diseases: principles, mechanisms of action, and future prospects. *Applications of Environmental Microbiology,* 71:4951-59.

De Weert, S., Kuiper, I., Kamilova, F., Mulders, I.H.M., Bloemberg, G.V., *et al.* 2007. The role of competitive root tip colonization in the biological control of tomato foot and root rot. In: *Biological Control of Plant Diseases,* (eds. S.B. Chincolkar and K.G. Mukerji), Haworth, London. pp. 103-122.

De Weert, S., Vermeiren, H., Mulders, I.H.M., Kuiper, I., Hendrickx, N., *et al.* 2002. Flagella-driven chemotoxis towards exudate components is an important trait for tomato root colonization by *Pseudomonas fluorescens. Molecular Plant Microbe Interactions,* 15:1173-1180.

Dey, R., Pal, K.K. and Tilak, K.V.B.R. 2004. Interaction of plant growth promoting rhizobacteria and AM fungi: Present status and future prospects. In: *Basic Research and Applicationsof Mycorrhizae* (eds. G.K. Podila and A.K. Varma). I.K. International Pvt. Ltd., New Delhi, India. pp. 83-110.

Dey, R., Pal, K.K., Bhatt D.M. and Chauhan, S.M. 2004. Growth promotion and yield enhancement of peanut (*Arachis hypogaea* L.) by application of plant growth promoting rhizobacteria. *Microbiological Research,* 159: 371-394.

Dobbelaere, S., Vanderleyden, J. and Okon, Y. 2003. Plant growth promoting effects on diazotrophs in the rhizosphere. *Critical Review of Plant Science,* 22:107-149.

Ebel, J. 1986. Phytoalexin synthesis: The biochemical analysis of the induction process. *Annual Review of Phytopathology,* 24:235-264.

Glick, B.R. 2003. Plant growth promoting bacteria. In *Moleclar Biology-Principles and Applications of Recombinant DNA* (eds. B.R. Glick and J.J. Pasternak), ASM Press, Washington DC, USA, pp. 436-454.

Glick, B.R., Cheng, Z., Czarny, J. and Duan, J. 2007. Promotion of plant growth by ACC deaminase-producing soil bacteria. *Europeon Journal of Plant Pathology,* 119:329-339.

Kerr, A. 1982. Biological control of soil-borne microbial pathogens and nematodes. In: *Advances in Agricultural Microbiology* (ed. N.S.Subba Roa). Oxford & IBH Pub. Co., New Delhi, India. pp.429-464.

Kloepper, J.W., Leong, J., Teintze, M. and Schroth, M.N. 1980. Enhanced plant growth by siderophores produced by plant-growth promoting rhizobacteria. *Nature,* 286: 885-886.

Kloepper, J.W., Ryu, C.M. and Zhang, S. 2004. Induced systemic resistance and promotion of plant growth by *Bacillus* spp. *Phytopathology,* 94:1259-1266.

Kosuge, T. and Nester, E.W. 1984. *Plant Microbe Interaction: Molecular and Genetic Perspectives.* Vol.1. Macmillan, New York, USA.

Kravchenko, L.V., Azarova, T.S., Makarova, N.M. and Tikhonovich, I.A. 2004. The effect of tryptophan present in plant root exudates on the phytostimulating activity of rhizobacteria. *Microbiology*, 73:156-158.

Leong, J. 1986. Siderophores: their biochemistry and possible role in the biocontrol of plant pathogens. *Annual Review of Phytopathology*, 24:187-209.

Lugtenberg, B. and Kamilova, F. 2009. Plant Growth Promoting Rhizobacteria. *Annual Review of Microbiology*. 63:541-556.

Lugtenberg, B.J.J. and Dekkers, L.C. 1999. What makes *Pseudomonas* bacteria rhizosphere competent ? *Environmental Microbiology*, 1:9-13.

Lynch, J.M.1982. The rhizosphere. In: *Experimental Microbial Ecology* (eds. R.G.Burns and J.H.Slater). Blackwell Scientific, Oxford, UK. pp.395-410.

Mavrodi, D.V., Blankenfeldt, W. and Thomashow, L.S. 2006. Phenazine compounds in fluorescent *Pseudomonas* spp. Biosynthesis and regulation. *Annual Review of Phytopathology*, 144:417-445.

Mukherji, K.G. and Subba Rao, N.S. 1982. Plant surface microflora and plant nutrition. In: *Advances in Agricultural Microbiology* (ed. N.S.Subba Rao). Oxford & IBH Pub.Co., New Delhi, India.

Nakas, J.P. and Hagedorn, C. (eds.) 1990. *Biotechnology of Plant Microbe Interactions*. McGraw-Hill Publishing Co., New York, USA.

Pal, K.K., Dey, R. and Bhatt, D.M. 2004. Groundnut (*Arachis hypogaea* L.) growth, yield and nutrient uptake as influenced by inoculation of plant growth promoting rhizobacteria *Journal of Oilseeds Research*, 21: 284-287.

Pandey, K.K. and Upadhyay, J.P. 2000. Microbial population from rhizosphere and non-rhizosphere soil of pigean pea: screening of resident antagonists and mode of parasitism. *Journal of Mycology and Plant Pathology*, 30:7-10.

Pliego, C., De Weert, S., Lamers, G., De Vicente, A., Bloemberg, G., *et al.* 2008. Two similar enhanced root-colonizing *Pseudomonas* strains differ largely in their colonization strategies of avocado roots and *Rosellinia neatrix* hyphae. *Environmental Microbiology*, 10:3295-3304.

Rodriguez, H., Mendoza, A., Antonia Cruz, M., Holgiun, G., Glick, B.R. and Bashan,Y.2006. Pleiotropic physiological effects in the plant growth promoting bacterium *Azospirillium brasilense* following chromosomal labeling in the *clp X* gene. *FEMS Microbial Ecology*, 57:217-225.

Ryu, C.M., Farag, M.A., Hu, C.H., Reddy, M.S., Wie, H.X., *et al.* 2003. Bacterial volatiles promote growth of *Arabidopsis*. *Proceedings of National Academy of Sciences*, USA, 100:4927- 4932.

Staley, T.E. and Brauer, D.K. 2006. Survival of a genetically modified root colonizing pseudomonad and *Rhizobium* strain in an acidic soil. *Soil Society of America Journal*, 70:1906-1913.

Subba Rao, N.S. and Dommergues, Y.R. (eds.) 1998. *Microbial Interaction in Agriculture and Forestry*. Vol.1. Oxford & IBH Pub. Co., New Delhi, India.

Tate, R.L.III. 2000. *Soil Microbiology*. John Wiley & Sons, New York, USA.

Tilak, K.V.B.R. and Srinivasa Reddy, B. 2006. *Bacillus cereus* and *B. circulans* - novel inoculants for crops. *Current Science*, 90: 642-644.

Tilak, K.V.B.R., Ranganayaki, N. and Manoharachari, C. 2006. Synergistic effects of plant growth promoting rhizobacteria and *Rhizobium* on nodulation and nitrogen fixation by pigeonpea. *Europeon Journal of Soil Science*, 57: 67-71.

Tilak, K.V.B.R., Pal, K.K. and Dey, R. 2010. *Microbes for Sustainable Agriculture*. I.K. International Publishing House Pvt. Ltd., New Delhi, India.

Tilak, K.V.B.R., Ranganayaki, N., Pal, K.K., Dey, R., Saxena, A.K., Nautial, C.S., Mittal, S., Tripathi, A.K. and Johri, B.N. 2005. Diversity of plant growth and soil health supporting bacteria. *Current Science*, 89:136-150.

Uren, N.C. 2007. Types, amounts, and possible functions of compounds released into the rhizosphere by soil-grown plants. In: *The Rhizosphere. Biochemistry and Organic Substances at the Soil-Plant Interface*, 2nd ed. (eds. R. Pinton, Z. Varanini and P. Nannipieri), CRC Press/ Taylor & Francis Group. Boca Raton, USA, pp. 1-21.

Van Loon, L.C. 2007. Plant responses to plant growth-promoting bacteria. *European Journal of Plant Pathology*, 119:243-254.

Vassilev, N., Vassileva, M. and Nicolaeva, I. 2006. Simultaneous P-solubilizing and biocontrol activity of microorganisms: potentials and future trends. *Applications of Microbiology and Biotechnology*, 71:137-144.

Vessey, J.K. (2003). Plant growth promoting rhizobacteria as biofertilizer. *Plant and Soil*, 255: 571-586.

Chapter-6

Microorganisms for the Biocontrol/Management of Soil-Borne Plant Diseases

Emphasis of present day agriculture is to produce more with lesser land, water and man power. Apart from natural calamities like drought, flood etc., biotic stresses like pests, diseases and weeds also influence the crop productivity greatly. Sixty to eighty per cent of the total crop losses are attributed to the devastations caused by pests and diseases. Farmer's income in recent years is decreasing gradually due to increased cost of inputs particularly in plant protection. The excessive dependence on chemical pesticides leads to the development of resistance in insect pests and pathogens, out breaks of secondary pests and pathogens / biotypes and occurrence of residues in food chain. To avoid this situation the concept of Integrated Pest Management (IPM)/ Integrated Disease Management (IDM) approach should be considered on top priority.

Several pathogens have been reported to cause serious diseases in many crop plants and trees. Various control measures including fungicides, tolerant cultivars, organic amendments, biocontrol agents and modified cultural practices have been tried for the management of the diseases. Management of soil-borne pathogens like, *Rhizoctonia solani, Sclerotium rolfsii, Fusarium oxysporum, Macrophomina phaseolina, Sclerotinia* spp., *Pythium* spp., *Phytophthora* spp. etc. using chemicals has been reported with limited success, since these pathogens can survive under a wide range of conditions, and it becomes very difficult to treat large volumes of soil / plant chemically to eliminate or reduce the pathogen population. The chemical control also has other limitations such as high cost, low cost benefit ratio, poor availability, selectivity, temporary effect, efficacy affected by physico-chemical and biological factors, development of disease resistance, resurgence of diseases, pollution of foods and feeds, health hazards, toxicity towards plants, animals and natural enemies, environmental pollution, etc.

Considering these limitations, there has been a growing awareness to develop such management practices, which alone or in combination with other practices could bring about a reasonably good degree of reduction of disease incidence and at the same time ensure sustainability of production, cost effectiveness and healthy ecosystem. Biological control is an important approach in this direction. It refers to "reduction of the amount of inoculum or disease producing activity of a pathogen accomplished by or through one or more organisms other than man".

There are different approaches for biological control of soil-borne plant pathogens. Nature always tries to maintain an equilibrium among the soil inhabitants. In any soil we can expect some microorganisms which may be antagonistic to plant pathogenic fungi. There are innumerable examples where biological control has been achieved by carefully manipulating the conditions which favour the growth and development of such native antagonists. Another approach is to introduce certain selected biocontrol agents to the soil.

Biological Control by Activating Native Organisms

The activity of native antagonists can be intensified during the period when control is needed. It can be accomplished by following different methods.

Cropping system

Each specific cropping system offers unique ecological niche for soil microorganisms. Some of these organisms will be beneficial while others may be detrimental to plant growth. Crop rotation is an established practice to control soil-borne diseases. Inspite of its wide host range, *Rhizoctonia solani* was subject to amelioration by crop rotation. The incidence of *Rhizoctonia* root rot of sugar beet was 65%

following wheat and sorghum, 35% after alfalfa (lucerne) and 15% after cotton, sunflower and fallow. Similarly the root rot of wheat caused by *Bipolaris sorokiniana* was influenced by crop rotation. After 2 years of wheat, the number of spores in soil had increased by 98% and with barley 206%. The number decreased by 36-46% following safflower, chickpea and cocksfoot which are non-host crops.

Manipulation of soil environment

The soil environment greatly affects the plant pathogen-antagonist interaction. Environment in biological control is operated by increasing the ability of the host to resist, tolerate or escape the pathogen; through increasing the capacity of the antagonist as a competitor, parasite or antibiotic producer or through weakening the ability of the pathogen to affect the host or resist the antagonist. Hence, understanding the conditions favouring the antagonists or pathogen will help to minimize the disease. It is well established that adverse growth conditions like inadequate water, unfavourable temperature, poor nutrient status etc. make plant more susceptible to the pathogen, while certain environmental conditions directly affect the pathogen.

The biocontrol of *Sclerotium rolfsii* can be achieved by first drying the field and then irrigating to keep it moist for two weeks or more and then sowing the crop. Germination of the sclerotia in *Rhizoctonia solani* was affected by alternate wetting and drying. A marked reduction in germination was observed after the seventh wetting and drying. The number of hyphae produced by the sclerotium gradually reduced after each germination.

Agrochemicals

Application of fertilizer or other agrochemicals could reduce the incidence of several root diseases. Some root diseases reduce and others become more severe when specific forms of nitrogen fertilizer are used. Diseases caused by *Fusarium, Rhizoctonia* and *Aphanomyces* increased in severity by ammonium nitrogen and decreased by the nitrate form. Application of ammonium sulphate probably increases the activity of *Trichoderma* spp. resulting in pathogen suppressiveness; and a shift in the remaining microbiota to a population that was more antagonistic to *Gaeumannomyces graminis tritici* than to *Trichoderma*. Application of $CaCO_3$ to a damping off conducive soil reduced the disease intensity in cucumber (caused by *Pseudomonas spendens*) from 78% to 30%.

Tillage

The method of tillage modifies soil structure, crumb size, aeration, temperature and water potential. It can change the soil environment and the biological balance to control root pathogen. Subsoil tillage to break a hardpan at the plough sole reduced *Fusarium* damage by permitting better root penetration. Tillage reduced the incidence of diseases like take-all of wheat, *Rhizoctonia* root rot, and *Pythium* root rot compared to no tillage / undisturbed soils.

Disease suppressive soils

Certain soils have a natural ability to suppress the disease. In these soils disease development is suppressed when the pathogen is present or is introduced in the presence of a susceptible host. Disease suppressive soils have been reported for *Gaeumannomyces graminis tritici, Phytophthora cinnamomi, Rhizoctonia solani* and *Pythium* spp. Severity of the take-all disease of wheat reached the maximum after 2-3 years of wheat monoculture and subsequently declined. This is referred to as take-all decline. This decline was due to the development of a suppressive factor, biological in nature. A violet pigmented antagonistic *Pseudomonas* was fostered by the suppressive soil.

Biological Control by Addition of Specific Antagonists

Addition of specific antagonists to the soil has attracted attention of scientists as a possible method of effective control of root diseases. Bacteria, fungi and mycophagous nematodes have been tried mostly. Depending upon the crop, pathogen, soil condition etc. the biocontrol agents exhibited varying levels of protection. Microbial antagonism is the ecological association between organisms where one or more of the participants are harmed or its activities curtailed. Though biocontrol agents include all classes/groups of organisms existing in an ecosystem, maximum emphasis for developing biocontrol programmes was given to fungal and bacterial biocontrol agents primarily because of ease of their mass multiplication and formulation. In recent years successful use of fungal biocontrol agents like *Trichoderma* spp. and *Gliocladium* spp. for the control of soil-borne diseases in several crops have been reported. Among the bacterial antagonists, the fluorescent pseudomonads are under intensive research because of their wide natural occurrence, biocontrol potential against fungal and nematode diseases as well as host defense inducing and plant growth promoting activities. The biocontrol agents provide protection against plant diseases either by direct action against the pathogen by mycoparasitism or antibiosis and indirectly by competing with the pathogen for nutrients, oxygen or space. Therefore, biological control of soil-borne pathogens using antagonistic organisms has become very much essential in recent times.

Microbial antagonists are living organisms which are highly dependent on environmental conditions. These organisms are expected to be very specific in suppressing other living organisms, such as plant pathogens, in their typical habitat. Contrary to pathogens, antagonists have to prove their efficiency in the habitat where they sometimes do not live or do not reside continuously. Therefore, antagonists are normally less adapted to the environment to which they are applied in order to control the pathogen. Antagonists introduced into the new habitat also have to compete against all organisms already present. Under these circumstances, they have to establish a high population density for a time period as long as possible to provide the expected protection.

The ideal antagonist should, therefore, colonize this new habitat very quickly to strongly limit the availability of niches for the pathogen. Additionally the antagonists should compete successfully against all those inhabitants that are neither pathogens nor antagonists and should be resistant to light, dryness and extreme temperature. These demands are better fulfilled more by fungi as compared to bacteria.

Fungi

Fungi are very common inhabitants of a wide range of natural habitats. They are adapted to the life cycle of crops and to varying environmental conditions, such as humidity and availability of space. They form specific structures for survival under unfavorable conditions for growth and reproduction. Additionally, fungi have a range of antagonistic modes of action, depending on the species. This gives the possibility to choose the mode of action according to the biology of the pathogen and other factors having an influence on effectiveness. However, important weaknesses of fungi that hinder their wide spread use are their insufficient competence to establish on the plant surface and their typically slow growth rates on this habitat. Regarding the necessity to produce large quantities of a potential biological control agent, fungi seem to be highly suitable. They can be mass cultivated by different techniques and on varying culture media. Formulation techniques are already in use for different requirements of fungal antagonists and environmental conditions.

Fungi already investigated in this respect that have shown a certain level of antagonistic activity include *Trichoderma* spp., *Gliocladium* spp., *Chaetomium globosum* and *Myrothecium verrucaria.* Efficacy of fungal species varied largely and, although many were promising in green house tests, this did not prove commercially effective. A reduction of grey mold in strawberry by 42% for *Cladosporium herbarum* and by 31 % for *Aureobasidium (Pullularia) pullulans* when applied during flowering in the green house has been reported. In field trials, effectiveness was confirmed only for *Aureobasidium pullulans.* Some workers have reported fungal and bacterial antagonists for biological control of grey mould in strawberry was very effective under field trials. Promising results were also reported by screening *Trichoderma* spp. for effectiveness against *Botrytis cinerea* on strawberry. A reduction of grey mould by 21-42% in field trials was observed.

Of the different fungi investigated for the control of soil-borne plant pathogens, species of *Trichoderma* proved to be the best antagonists. The different species of *Trichoderma* investigated for the biocontrol potential are *T. viride, T. harzianum, T. koningii, T. virens, T. hamatum, T. polysporum, T. aureoviride, T. longibrachiatum* and *T. pseudokoningii.* Of these species *T. harzianum* and *T. viride* were found to be effective in controlling several soil-borne plant diseases like root rot of apple, citrus, chilli, brinjal, french bean, groundnut, safflower, sesamum, sunflower, cotton, sugarcane, chick pea, betelvine, casuarina etc.

Arbuscular mycorrhizal fungi are geographically ubiquitous and occur over a board ecological range. They are commonly found in association with agricultural and horticultural crops, most tropical tree species and some temperate tree species. Enhanced growth of plants because of AM fungal inoculation is well documented. The improved plant growth is attributed to increased uptake of diffusion-limited nutrients like P, Zn, Cu etc., production of plant hormones and improved resistance against abiotic and biotic stresses.

Most of the studies on AM fungi-root pathogens interaction suggest that AM fungi decreased or mitigated the disease severity. Consistent reduction of disease symptoms has been described for fungal pathogens such as *Phytophthora parasitica, P. cactorum, P. vignae, Gaeumannomyces graminis* var. *tritici, Fusarium oxysporum, Chalara*

(Thielaviopsis) basicola, Rhizoctonia solani, R. bataticola, Sclerotium rolfsii, Pythium ultimum, P. splendens, Dothiorella gregania, Botrytis fabae, Ganoderma pseudoferreum and *Aphanomyces* spp.; bacteria such as *Pseudomonas syringae* and *Ralstonia solanacearum* and nematodes such as *Meloidogyne avenaria, M. incognita, M. hapla, M. javanica, Tylenchulus semipenetrans, T. vulgaris, Pratylenchus brachyurus* and *Radopholus similis.*

Studies conducted so far suggest that the mechanisms of suppression may be due to morphological, physiological and biological alterations in the host. It has been demonstrated that AM fungal colonization induces remarkable changes in root system morphology, as well as in the meristematic and nuclear activities of root cells. This might affect rhizosphere interactions and particularly pathogen-infection development. The most frequent consequence of AM fungal colonization is an increase in branching, resulting in a relatively larger proportion of higher order roots in the root system. However, the significance of this finding for plant protection has not yet been sufficiently considered. Thickening of the cell walls through lignification and production of other polysaccharides in mycorrhizal plants preventing penetration and growth of pathogens like *Fusarium oxysporum* and *Phoma terrestris* have been demonstrated. A stronger vascular system observed in mycorrhizal plants will increase the flow of nutrients, impart greater mechanical strength and diminish the effect of vascular pathogens. Working with AM-*Phytophthora* interaction it was observed that the pathogen does not penetrate cortical cells containing arbuscule; suggesting that localized competition for infection / colonization site does occur between the pathogen and the AM fungus.

Colonization of a plant by AM fungi alters the host physiology and in turn the root exudation pattern. Decreased root exudation in mycorrhizal plants, because of increased membrane phopholipid content, possibly helps in reducing the infection by root pathogens. Higher concentration of ortho-dihydroxy phenols present in mycorrhizal plants compared to non-mycorrhizal plants was found to be inhibitory to the root rot pathogen *Sclerotium rolfsii*. The activation of specific plant defence mechanisms as a response to AM fungal colonization is an obvious basis for the protective capacity of AM fungi. Among the compounds involved in plant defence studied in relationship to AM formation are phytoalexins, chitinases, β-1,3-

glucanase, peroxidases, pathogenesis related (PR) proteins, hydroxyproline-rich glycoproteins (HRGP) and phenolics.

Mycorrhizal plants harbour higher population of microorganisms in the rhizosphere thus making it difficult for the pathogen to compete and gain access to the root. Microorganisms producing siderophores, which are low molecular weight chelating agents that have high affinity for ferric iron and thus fungistatic to many pathogens, were observed in higher numbers in the rhizosphere of mycorrhizal plants. Mycorrhizal plants harbouring more actinomycetes antagonistic to root pathogens has been reported.

Co-inoculation of PGPRs along with AM fungi protecting the plants better than AM fungus alone against root pathogens has been reported recently by many workers. Wilt of the medicinal plant *Coleus forskohlii* caused by *Fusarium chlamydosporum* is very serious in India. Inoculation with AM fungus plus *Trichoderma viride* was found to increase root yield and root forskolin concentration, and reduce the severity of the disease significantly under field conditions. (Table 18).

Table 18: Influence of biocontrol agents on the root yield, root forskolin concentration and plant disease index of *Coleus forskolii*

Treatment	Root dry weight (g/plant)	Root forskolin concentration (%)	Plant disease index
Control	0.03	0.30	85.55
Glomus mosseae	0.12	0.37	68.19
Trichoderma viride	0.17	0.34	65.69
Pseudomonas fluorescens	0.12	0.35	71.52
Emisan	0.10	0.32	79.74
G. mosseae + T.viride	0.27	0.55	33.28
G. mosseae + P. fluorescens	0.18	0.38	56.51
P. fluorescens + T. viride	0.23	0.39	54.79

Most of the AM-root pathogen interaction studies have been conducted in crop plants important in agriculture and horticulture. But the information available on forest tree species is scanty. There are large number of root pathogens causing diseases of forest tree species, especially in the nursery. Availability of healthy stock of seedlings for planting in the field is an important aspect of forest management. Hence there is a need to carry out more studies on the role of AM fungi in the control of root pathogens of forest tree species. Mycorrhizal technology can thus play an important role in production of low cost quality seedlings for afforestation and reforestation programmes in forestry.

Bacteria

From the biological control perspective, bacteria are very promising candidates. However, bacteria are very susceptible to unfavourable environmental conditions. Humidity is an important factor influencing bacterial control agents, since bacterial cells need a high degree of water availability for survival and growth. Temperature is less important regarding survival of bacteria, but it is important for biosynthesis of suppressive substances such as antibiotics.

The use of antagonistic bacteria seems to be more promising when applied against bacterial rather than fungal pathogens, since both pathogen and antagonist have very similar environmental requirements. The mode of action of bacteria can be antibiosis, due to the production of antibiotic substances or enzymes, competition for nutrients or space or plant growth promotion. These antibiotics may kill the attacked organism or they may suppress different stages in the life cycle of the organism, such as spore germination, elongation of germ tubes or mycelial growth. Enzymes are very important in the context of parasitism, but they have also been reported for bacterial antagonists not acting as parasites. Enzymes degrade the cell wall of affected organisms leading to the efflux of cytoplasm or disturbance of normal living processes. Bacteria may successfully compete for nutrients in all habitats that they are able to colonize. On aerial plant parts, they are known as early colonizers competing mainly for amino acids. A very specific mechanism of competition has been described for fluorescent pseudomonads, for example *Pseudomonas fluorescens,* since they are able to form siderophores, specific binding sites for Fe^{3+} ions that are able to fix iron, there by reducing its availability for other organisms.

Potential biocontrol agents, fungi and bacteria, for the management of diseases are given in table 19.

Table 19 : Biocontrol agents for the management of diseases of crop plants in India

Crop	Disease / Pathogen	Possible Biological Control Agent
Rice	Bacterial leaf blight (*Xanthomonas oryzae*)	*Bacillus* spp.
Wheat	Loose smut (*Ustilago segetum*)	*Trichoderma koningii*
Maize	Charcoal rot (*Macrophomina phaseolina*)	*T. viride*
Pigeon pea	Wilt (*Fusarium udum*)	*T. viride / T. harzianum*
Chickpea	Wilt (*Fusarium oxysporum* f.sp. *ciceri*)	*T. virens / T. harzianum*
	Collar rot (*Sclerotium rolfsii*)	*T. harzianum / T. viride*
Ground nut	Stem rot	*T. harzianum* (*Sclerotium rolfsii*)
Brinjal	Damping-off (*Phytophthora* or *Pythium* sp. or *Fusarium* sp.)	*T. harzianum / T. viride*
Cabbage	Damping-off (*Rhizoctonia solani*)	*T. harzianum*
Potato	Bacterial wilt (*Ralstonia solanacearum*)	*Bacillus cereus / B. subtilis*
Tomato	Damping off and wilt (*Pythium indicum, F. oxysporum* f. sp. *lycopersici*)	*T. viride, T. harzianum*
Black pepper	Foot rot and root rot (*Phythophthora capsici*)	*T. harzianum* / AM fungi
Cardomom	Capsule rot (*Phythophthora meadii, P. nicotianae var. nicotianae*)	*T. harzianum*
Ginger	Rhizome rot (*Pythium aphanidermatum*)	*T. harzianum*
Banana	Panama disease (*Fusarium oxysporum* f.sp. *cubense*)	*Pseudomonas fluorescens / T. viride*

Cotton	Root rot (*Rhizoctonia* sp. / *M. phaseolina*)	*T. viride, T. harzianum, P. fluorescens*
Sugarcane	Red rot (*Colletotrichum falcatum*)	*P. fluorescens* VPT4, ARRIG and EPI and *P. putida* KKM1

Mode of Action of Biocontrol Agents

Biological control by antagonistic organisms uses naturally occurring mechanisms to suppress plant pathogens. The modes of action are competition for an ecological niche or substrate, production of inhibitory allelo-chemicals, parasitism and induction of systemic resistance (ISR) in host plants to a broad spectrum of pathogens and/ or abiotic stresses.

Competition for an ecological niche or a substrate

Rhizosphere competence

Despite their potential as low-input practical agents of plant protection, application of biocontrol agents has been hampered by inconsistent performance in the field tests; this is usually attributed to their poor rhizosphere competence. Rhizosphere competence of biocontrol agents comprises effective root colonization combined with the ability to survive and proliferate along growing plant roots over a considerable time period, in the presence of the indigenous microflora. The root surface and surrounding rhizosphere are significant carbon sinks. Photosynthate allocation to this zone can be as high as 40%. Thus, along root surfaces there are various suitable nutrient rich niches attracting a great diversity of microorganisms, including phytopathogens. Competition for these nutrients and niches is a fundamental mechanism by which microorganisms protect plants from phytopathogens.

Since root exudates are the primary source of nutrients for rhizosphere microorganisms, rhizosphere competence implies that biocontrol organisms are well adapted to their utilization. Despite the fact that sugars have often been reported as the major carbon source in exudates, the ability to use specific sugars does not play a major role in tomato root colonization. Similarly, although amino acids are present in root exudates, the bioavailability of amino acids alone is considered insufficient to support root tip colonization by auxotropic mutants of *P. fluorescens* WCS365. In contrast, it has been reported that amino acid synthesis is required for root colonization by *P.*

fluorescens WCS365, indicating that amino acid prototrophy is involved in rhizosphere competence.

Competition for nutrients

Outcompeting pathogens for nutrients is a biocontrol strategy that can be used against pathogens, where the biocontrol agent preferentially uses external nutrients before using the host plant. Pathogens need external nutrients for conidial germination, for germ tube elongation or for formation of infection structures. Due to this dependency on external nutrients, they are in a very weak position in the phase between arrival and penetration of the host. Yeasts and bacteria are strong competitors for nutrients due to their relatively large surface area as compared to hyphae and are, therefore, often used as biocontrol agents. In addition to the general competitiveness of bacteria for nutrients, the very specific mechanism of fluorescent pseudomonads should once more be mentioned. These bacteria are able to form siderophores, which are specific binding site for iron3^{+} ions. In the presence of fluorescent pseudomonads, for example, *P. fluorescens,* availability of iron3^{+} is strongly restricted to other organisms, resulting in their reduced growth and multiplication.

Botrytis cinerea is one of those pathogens requiring external nutrients for conidial germination and germ tube elongation. Therefore, competition for nutrients strongly affects conidial germination of this pathogen, especially under environmental conditions of low nutrient availability. These findings are supported by results from Brodie and Blakeman (1975) who reported nearly complete suppression of conidial germination of *B. cinerea* by *Pseudomonas* spp. The bacteria successfully competed for glutamines added to the suspension and exhibited an uptake of 85 % of the nutrients as compared to only 10 % taken up by conidia of *B. cinerea.* Bacteria took up most of the nutrients within a short period of 3 hours due to a significant increase in physiological activity. Similar results were obtained for yeasts when sugar uptake was investigated. Yeasts strongly competed for sugars by taking them up faster and in higher quantity as compared to conidia of *B. cinerea.*

Competition between microbes for nutrients is a phenomenon, which also occurs under natural conditions on leaf surfaces. Yeasts and bacteria live as saprophytes on the phylloplane and are able to establish substantial populations in the presence of small quantities of nutrients if sufficient water is available. Nutrients originate from honeydew, pollen and many plant components, and are normally sufficiently available to support conidial germination. If a mixed antagonist population, containing bacteria and yeasts, is present, amino acids are taken up by bacteria, whereas yeast take up both sugars and amino acids.

Competition for space

Living organisms in nature also commonly compete for space. Each organism has specific requirements in the habitat and, therefore, prefers specific niches for living or survival. The environmental factors, which determine characteristics of niche, are temperature, humidity, the nature of plant exudates, plant surface characteristics, amount of UV radiation, as well as other factors. For plant pathogens, penetration sites for infection and niche for survival might be very important. A special case of competition for space is the struggle for infection sites, which may occur if antagonists are able to occupy the specific places where recognition mechanisms between host plant and pathogen take place. If these places are no more available for pathogens, the necessary procedure of recognition cannot take place and infection does not occur. A well known interaction of this type is the competition between *Agrobacterium tumefaciens,* the casual agent of crown gall in many crops and the antagonist *Agrobacterium radiobacter* K84. One mode of action described for *A. radiobacter* is the competition for binding sites at the host plant.

Bacterial traits required for effective root colonization are subject to phase variation, a regulatory process for DNA rearrangements orchestrated by site specific recombinase. In certain bacteria, efficient root colonization is linked to their ability to secrete a site specific recombinase. Transfer of the site specific recombinase gene from a rhizosphere competent *P. fluorescens* into a rhizosphere incompetent *Pseudomonas* strain enhanced its ability to colonize root tips.

Production of inhibitory allelochemicals

Competition for iron and the role of siderophores

Iron is an essential growth element for all living organisms. The scarcity of bioavailable iron in soil habitats and on plant surfaces foments a furious competition. Under iron limiting conditions some biocontrol organisms produce low molecular weight compounds called siderophores to competitively acquire ferric ion. Siderophores are secondary metabolites of low molecular weight secreted by some organisms which under low iron conditions, selectively chelate iron making it unavailable to other deleterious organisms like soil-borne pathogens thus reducing their population. Although various bacterial siderophores differ in their abilities to sequester iron, in general, they deprive pathogenic fungi of this essential element since the fungal siderophores have lower affinity. Some biocontrol organisms go one step further and draw iron from heterologous siderophores produced by cohabiting microorganisms. Siderophore mediated antagonism is the method of biological control of *Phytophthora capsici* by *Pseudomonas fluorescens.*

Antibiosis

Antibiosis is caused by suppressive substances (antibiotics) produced by living organisms to defend a habitat or to colonize a new one. These substances, in very low concentrations, are able to suppress growth of microorganisms and the effect may cause death or otherwise interfere in the biological processes. Bacteriocins represent a special group, which are very similar to antibiotics and have an antibiotic effect on related organisms. The most predominant example is agrocin 84 produced by the antagonist *A. radiobacter.* This antagonist is highly effective in controlling *A. tumefaciens* and antibiosis seems to be its most important mode of action. Antibiosis may also be caused by enzymes that are able to transform a specific substrate into a toxic substance. One example for this is the enzyme glucose-oxidase that degrades glucose into hydrogen peroxide, which then causes antibiosis. Volatiles, such as hydrogen cyanide produced by rhizobacteria, represent the third group of agents, which can suppress other organisms. These volatiles may contribute to biological control of soil-borne plant pathogens.

The production of antibiotics can be detected using dual culture tests *in vitro,* consecutively exemplified with a bacterial antagonist and a fungal pathogen (Plate 17). Antibiotic substances produced by the bacterial colony on nutrient agar diffuse into the agar and cause an inhibition zone around the fungal colony. The reaction of the fungal colony to these antibiotics depends on several factors, such as the concentration quantity and effectiveness of the antibiotics as well as the susceptibility of the fungus. However, growth of the fungal colony is normally inhibited by these antibiotics present in the near vicinity of the antagonist colony, resulting in the already mentioned inhibition zone. In addition to climatic factors such as temperature and humidity, the production of antibiotics and enzymes may depend on the nutritional status of the environment or on the presence of key substances. For example, production of β-1, 3 glucanase by *Paenibacillus polymyxa* on potato agar, but not on potato dextrose agar (PDA) has been observed. Similarly *Bacillus pumilus* producing cellulase and mannase on potato agar, but not on PDA has also been reported. *B. subtilis* produced high quantities of suppressive substances causing large inhibition zones when cultured on cabbage extract agar as compared to nutrient broth agar.

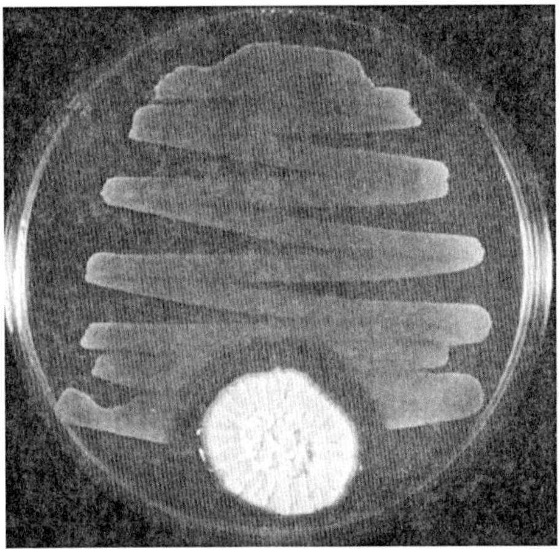

Plate 17 : Antagonistic bacterium showing zone of inhibition against a fungal pathogen

The basis of antibiosis as a biocontrol mechanism has become increasingly better understood over the past two decades. A variety of antibiotics have been identified, including compounds such as amphisin, 2, 4-diacetylphloroglucinol (DAPG), hydrogen cyanide, oomycin A, phenazine, pyoluteorin, pyrrolnitrin, tensin, tropolone, and cyclic lipopeptides produced by psedomonads and oligomycin A, kanosamine, zwittermicin A, and xanthobaccin produced by *Bacillus, Streptomyces, and Stenotrophomonas* spp. Interestingly, some antibiotics produced by biocontrol organisms are finding new uses as experimental pharmaceuticals, and this group of bacteria may offer an untapped resource for compounds to deal with the alarming ascent of multidrug-resistant human pathogenic bacteria.

Detoxification and degradation of virulence factors

Another mechanism of biological control is the detoxification of pathogen virulence factors. For example, certain biocontrol agents are able to detoxify albicidin toxin produced by *Xanthomonas albilineans*. The detoxification mechanisms include production of a protein that reversibly binds the toxin in both *Klebsiella oxytoca* and *Alcaligenes denitrificans*, as well as an irreversible detoxification of albicidin mediated by an esterase that occurs in *Pantoea dispersa*. Several different microorganisms, including strains of *Burkholderia cepacia*, can also hydrolyze fusaric acid, a phytotoxin produced by various *Fusarium* species.

Parasitism

Some biocontrol organisms parasitize the pathogen and destroy or reduce their population. This is possible due to production of specific cell wall degrading enzymes that allow the parasite to penetrate into the host or to completely or partly destroy host structures before feeding on them. In the case of fungi, penetration can be supported by previous coiling around host hyphae and production of specific penetration hyphae. The most prominent parasitic fungi are *Trichoderma* spp. that feed not only on microorganisms but also on insects and other organisms. *Trichoderma pseudokoningii*, for example, penetrates the hyphae of the host fungi and destroys them. The same mechanism has been reported for *Trichoderma viride*, which is strongly antagonistic and mycoparasitic to soil-borne *Phytophthora* spp. *in vitro*. One species that has gained high importance is *Coniothyrium minitans*, a parasite that is considered to be primarily a mycoparasite of sclerotia of *Sclerotinia* spp. After application of *Coniothyrium minitans* inoculum to the soil, mycelium grows to neighbouring sclerotia and the fungus feeds on them, resulting in destruction of the resting structures of the host. This parasitic fungus is characterized by its high host specificity and high potential to survive in the field.

Enzymes such as chitinases, β-1,3-glucanase and cellulases are very important for attacking the host cell wall in order to allow subsequent penetration of host structures. Production of these enzymes depends on the nutrient source *i.e.* the host cell wall that has to be attacked. Screening several *Trichoderma* isolates, it was found that those isolates which produced high levels of chitinases and β-1,3-glucanase were effective against *Rhizoctonia solani* and *Sclerotium rolfisii*, but not against *Pythium aphanidermatum*, because for this pathogen cellulases are required for attack. It has also been reported that protease was the most important enzyme produced by *T. harzianum* isolate T39 when used for biocontrol of *Botrytis cinerea*. Protease has been detected in culture liquid and effectively inhibited conidial germination and germ tube elongation of the pathogen *in vitro* and on leaf surface. The ability to produce extracellular chitinases is considered crucial for *Serratia marcescens* to act as antagonist against *Sclerotium rolfsii,* and for *Paenibacillus* spp. strain 300 and *Streptomyces* spp. strain 385 to suppress *Fusarium oxysporum* f.sp. *cucumerinum.* It has been also demonstrated that extracellular chitinase and laminarinase synthesized by *Pseudomonas stutzeri* digest and lyse mycelia of *Fusarium solani.*

Biopriming plants with some bacteria can also provide systemic resistance against a broad spectrum of plant pathogens. Diseases of fungal, bacterial and viral origin, can be reduced after its application. Certain bacteria trigger a phenomenon known as induced systemic resistance (ISR) phenotypically similar to systemic acquired resistance (SAR). SAR develops when plants successfully activate their defense mechanism in response to primary infection by a pathogen notably when the latter induces a hypersensitive reaction through which it becomes limited in a local necrotic lesion of brown, desiccated tissue. As SAR, ISR is effective against different types of pathogens but differs from SAR in that the inducing bacterium does not cause visible symptoms on the host plant. Plant growth promoting rhizobacteria - elicited ISR was first observed on carnation (*Dianthus caryophillus*) with reduced susceptibility to wilt caused by *Fusarium* spp. and on cucumber (*Cucumis sativus*) with reduced susceptibility to foliar disease caused by *Colletotrichum orbiculare.* Manifestation of ISR is dependent on the combination of host plant and bacterial strain. Most reports of bacteria mediated ISR involve free living rhizobacterial strains, but endophytic bacteria have also been observed to have ISR activity. For example, ISR was triggered by *P. fluorescens* EPI against red rot caused by *Colletotrichum falcatum* on sugarcane, *Burkholderia phytofirmans* PsJN against *Botrytis cinerea* on grapevine and *Verticillium dahliae* on tomato, *Pseudomonas denitrificans* 1-15 and *P. putida* 5-48 against *Ceratocystis fagacearum* on oak, *P. fluorescens* 63-28 against *F. oxysporum* f.sp. *radicis - lycopersici* on tomato and *Pythium ultimum* and *F. oxysporum* f.sp. *pisi* on pea roots and *Bacillus pumilus* SE34 against *F. oxysporum* f.sp. *pisi* on pea roots and *F. oxysporum* f.sp. *vasinfectum* on cotton roots.

Bacterial triggered ISR fortifies plant cell wall strength and alters host physiology and metabolic responses leading to an enhanced synthesis of plant defense chemicals upon challenge by pathogens and/ or abiotic stress factors. After inoculation of tomato with endophytic *P. fluorescens* WCS417r, a thickening of the outer tangential and outer most part of the radial side of the first layer of cortical cell walls occurred. In addition to structural barriers, physiological or biochemical changes can also be induced by biocontrol organisms inducing ISR such as accumulation of pathogenesis-related proteins (PR proteins), phenolic compounds, chitinases, peroxidases, phenylalanine ammonia lyase, phytoalexins, polyphenol oxidase and/ or chalcone synthase. Recent evidence indicates that induction of some of these plant defense compounds may be triggered by the same N-acyl homoserine lactones that bacteria use for intraspecific signaling.

A major distinction often drawn between ISR and SAR is the dependence of the latter on the accumulation of salicylic acid (SA). Some bacteria activating ISR do trigger an SA-dependent signaling pathway by producing nanogram amounts of SA in the rhizosphere. However, the majority of bacteria that activate ISR appear to do so via a SA-independent pathway involving jasmonate and ethylene signals. ISR is associated with an increase in sensitivity to these hormones rather than an increase in their production, which might lead to the activation of a partially different set of defense genes.

Methods of Application of Biocontrol Agents

Seed treatment

When seeds are treated with antagonists, the antagonists are present from the beginning of plant growth at the location where protection is required. Due to this early presence, antagonists are able to colonize the seed and the young roots successfully. Competition by other inhabitants of the rhizosphere is significantly reduced and antagonists are able to establish a high population density. *B. subtilis* and *Pseudomonas* spp. are the antagonists mainly investigated for this purpose. Seed treatment of potatoes with *B. subtilis* successfully reduced the incidence of *R. solani and Streptomyces scabies* in green house and field trials. Reduction of disease incidence in field trials varied between different experimental years but reached a level of up to 50 % for *R. solani* and up to 67 % for *S. scabies.*

The technique used for seed treatment depends on the type of seed to be treated. Very simple techniques, such as dipping of seeds, may be used since positive effects of such treatments with antagonists have been reported. However, the most suitable technique for small seeds is seed coating, including the introduction of antagonists into the coating substance. Coating of seeds has several advantages that increase the performance of antagonist application. Chemical substances used for coating can be chosen according to the requirements of the antagonists and can have several functions. They may serve as nutrients, as a protection barrier against physical, chemical or biological influence and for gradual release of antagonists.

Application of antagonists to the soil has several advantages. The most important one might be the relatively low variation that occurs in environmental conditions. Therefore, antagonists have less stress and higher probability of survival after application. Treatments of soil with antagonists may target the reduction of primary inoculum sources, for example, the destruction of hibernating sclerotia. *Trichoderma* spp. able to decompose sclerotia of *Botrytis cinerea* have been identified. They were able to decompose approximately 70 % of sclerotia within 60 days at 5°C. Another antagonistic fungus used the same way is *Coniothyrium minitans,* a parasite that attacks sclerotia of *S. sclerotiorum* and reduces primary inoculum in the soil. The antagonist is added to the soil and the mycelium grows to the sclerotia in the near vicinity in order to penetrate and decompose them. Due to destruction of sclerotia, disease incidence in the following susceptible crop is reduced. Several antagonists, especially bacteria, have been investigated for their potential to reduce infection by root invading pathogens. Most prominent examples might be *Pseudomonas* spp. (for example, *P. fluorescens, P. chlororaphis*) and *Bacillus* spp. (for example, *B. subtilis, B. cereus, B. megaterium, B. licheniformis*). These antagonists are suitable candidates for soil application.

Formulation of Antagonists

Formulation of antagonists should fulfill a variety of functions with the overarching goal to support viability and by this stabilize and increase performance of antagonists. Antagonists should, therefore, be treated carefully and supported optimally in all phases of this process, namely in processing, storage and use in practice. Besides improvement of storage ability, which should amount to at least one year at room temperature, formulation should also improve adhesion and distribution of cells on the plant surface. Formulation should be cheap, biodegradable and user friendly. Due to this wide and important range of functions, formulation is seen as a key factor for practical use of antagonists, because this converts the laboratory phenomenon "antagonist" into a commercial product. Alginate, xantham, cellulose, tween 80 etc. are commonly used as formulation substances. Encapsulated alginate beads with *Pseudomonas fluorescens* were able to colonize roots successfully as freshly produced bacteria. Colonization of roots was also very successful at a low antagonist inoculum concentration and when applying the antagonist product in the field.

Success Stories

Several microorganisms have shown potentialities as biological control agents against important plant diseases caused by soil-borne pathogens. They are *Pseudomonas fluorescens* against the take-all disease of wheat caused by *Gaeumannomyces graminis tritici, Erwinia carotovora* infection of wheat, *Thielaviopsis basicola* infection of tobacco and damping off caused by *Pythium* in cotton; *Bacillus subtilis* against *Fusarium roseum* wilt of corn; *Trichoderma harzianum* against *Alternaria* spp. infection of radish; *Penicillium oxalicum* against root rot of peas; non-pathogenic *Fusarium oxysporum* against *Fusarium* wilt of cucumber; *Trichoderma viride* against *Verticillium* wilt of tomatoes; *Cytophaga* spp. against damping off of conifer seedlings; *Bacillus* spp., *Penicillium* spp., and *Alcaligenes* spp. against crown gall disease of cherry seedlings caused by *Agrobacterium tumefaciens; Chaetomium globosum* against damping off of sugar beets; *Pseudomonas putida* against *Fusarium solani* wilt of beans; *Bacillus* spp. and *Pseudomonas* spp. against *Fusarium oxysporum* wilt of carnations.

Although research on biological control of plant diseases has been carried out for a number of years, it is relevant to mention some examples of success in this regard. One example is the use of *Peniophora gigantea* to control *Heterobasidium annosum* infection of pine trees. In the UK and the USA, felled pine tree stumps are coated with *Peniophora gigantea* which prevents the colonization of stumps by *Heterobasidium annosum,* thereby restricting the survival and multiplication of the pathogen. The second example is the use of *Agrobacterium radiobacter* in the control of crown gall disease caused by *Agrobacterium tumefaciens.* Considerable damage to stone fruit, pome fruit, other rosaceous plants, walnut, grapevine, chrysanthemum and dahlia is caused by crown gall infection (plant cancer). This disease has been successfully controlled in Australia by using the strain 84 of the non-pathogenic *Agrobacterium radiobacter,* which produces an unusual antibiotic called "Agrosin 84" which belongs to a group of antibiotic known as nucleotide bacteriocins. A bacteriocin is a substance produced by certain strains of bacteria and active on other strains of the same or closely related species.

A large conjugative plasmid called Ti plasmid (Tumour inducing circular DNA molecule) controls crown gall infection. This plasmid can be transferred to another cell by conjugation and in this process a non-pathogen can become a pathogen. The plasmid directs the synthesis of chemicals called opines-nopaline, agrocinopine A, octopine and agropine. Strains of *A. tumefaciens* which produce the first two opines are known as 'nopaline strains' and those which produce the other opines are known as 'octopine strains'. All octopine strains are not controlled by agrosin 84 but nopaline strains are susceptible to the bacteriocin and fortunately crown gall on stone fruit is caused by nopaline strains of *A. tumefaciens* and hence can be controlled.

Biological Control of Plant Parasitic Nematodes

Biological control aims in increasing the parasites and predators of nematodes in the soils, to increase the mortality of plant parasitic nematodes. Plant parasitic nematodes persisting in soil attack several crop plants and reduce their growth and yield. Estimated overall annual yield loss on the world's major crops due to damage by plant parasitic nematodes is around 13 %. Nematode control measures may either be corrective (e.g. chemical) or preventive (cultural) but the biological control has both these possibilities. Biological nematode control in relation to crop production system is a subject of considerable current interest, because of a perceived urgency to develop and adopt safe, economic and efficient method for managing nematode pests. The havoc caused by potato cyst eelworm, *Heterodera rostochiensis*; burrowing nematode, *Radopholus similis*; spreading decline of citrus, *Tylenchulus semipenetrans*; the interaction between nematodes and fungi causing plant disease complexes, the discovery that some of the nematodes are capable of transmitting viruses triggered interest in research on plant parasitic nematodes.

Among the best known plant parasitic nematodes are the root-knot, belonging to the genus *Meloidogyne*, which have a wide host range and are widely prevalent all over the world. These produce conspicuous galls on the roots and the infestation can be recognized easily in fields. The wide distribution of these nematodes through tropics, sub-tropics and temperate regions and their occurrence in cultivated fields and green house as well, make them as one of the most common agricultural pests. It is specially a serious problem with vegetable growers and also cotton and tobacco. Involvement of various ecto-parasitic nematodes that feed ectoparasitically on roots of the plants, along with endo-or semi-endoparasites, notably *Pratylenchus* sp., *Tylenchulus* sp. and *Rotylenchulus* sp., adds to the dimensions of losses caused by these to economically important crop plants. Even though a very large number of plant parasitic species have been recorded and documented, about 15 have been identified, as serious causing considerable damage in different parts of the world (Table 20).

Table 20 : Nematode diseases of high economic importance

Nematode	Crops attacked
Root-knot nematodes (*Meloidogyne* spp.) *M. incognita* *M. javanica* *M. hapla* *M. arenaria*	Vegetables, pulses, oil seeds, fruits, fibre plantation and few cereal crops
Reniform nematode (*Rotylenchulus reniformis*)	Vegetables, pulses, oilseeds, some fruits and fibre crops
Golden cyst nematode (*Globodera rostochiensis*)	Potato
Stem and Bulb nematode (*Ditylenchus dipsaci*)	Vegetables, cereals, ornamentals
Cereal root Eelworm (*Heterodera avenae*)	Wheat, barley, oats, rye, maize
Sugarbeet cyst nematode (*Heterodera schachtii*)	sugarbeet and many other crops
Burrowing nematodes (*Radopholus similis*)	Banana, pepper, coconut
Citrus nematode (*Tylenchulus semipenetrans*)	Grapefruit and oranges
Rice root nematodes (*Hirschmanniella oryzae*)	Rice
Pine wood nematode (*Bursaphelenchus xylophilus*)	Pine

Traditionally nematodes are controlled by several methods which include, regulatory methods, cultural, chemical, host resistance, physical, integrated and biological methods. Biological control includes the use of predaceous or parasitic organisms, such as fungi, bacteria, protozoa, viruses, collembolans, mites etc. in the management of nematodes by skillful manipulation of the biosphere against nematode pests for achieving maximum benefits without upsetting the environment. The effectiveness of biocontrol agents depends upon some desired specific characteristics *viz.* (i) easy culturability; (ii) environmental adaptability; (iii) high reproductive potential; (iv) longevity; (v) easy applicability; (vi) safety to non-target organisms; (vii) ecological compatibility; (viii) capacity to produce toxic metabolites; (ix) host specificity; (x) genetic stability; (xi) high nematode reducing potential, etc.

The role of potential biocontrol agents such as bacteria and fungi in reducing population of parasitic nematodes is discussed below.

Bacteria

Depending upon their mode of action, bacteria may be categorized mainly into two groups. The first group includes bacteria which release metabolites that have a killing or inhibitory effect on plant nematodes (*Bacillus, Azotobacter*). The second group includes bacteria which parasitise directly on nematodes, thus affecting the penetration and reproduction of nematodes (*Pasteuria*).

Bacillus thuringiensis: Various formulations of *B. thuringiensis* are toxic to eggs and larvae of *Meloidogyne incognita* in *in vitro* conditions. The exotoxin produced by *B. thuringiensis* was found toxic to the populations of *Meloidogyne*. It also deterred *M. incognita* juveniles forming galls on tomato roots. Dipel and SAN 415 strains of *B. thuringiensis* sub sp. *kurstaki* have been found to suppress populations of both *M. javanica* and *Tylenchulus semipenetrans*. SAN 415 was found to be as effective as the standard nematicide, Fenamiphos. CR 371, another isolate of *B. thuringiensis* was effective in reducing galling by root-knot nematode on tomato and also resulted in reduced population of *Rotylenchulus reniformis*.

Azotobacter chroococcum: The heterotrophic bacterium, *A. chroococcum* antagonistic to *M. incognita*, inhibited the hatching of egg masses and prevented the larvae to penetrate into the roots of brinjal to form galls.

Pasteuria penetrans: This bacterial obligate parasite of many nematodes has received greater attention in recent years. It was first described as a sporozoan, *Duboscqia penetrans*. Its prokaryotic and bacterial affinities were established later, and the name *Bacillus penetrans* was adopted. Later *B. penetrans* was renamed as *Pasteuria penetrans. P. penetrans* is known to parasitize 205 nematode species belonging to 96 genera from 10 orders of the phylum Nematoda. It occurs widely (recorded from 51 countries) and is a potential biocontrol agent. It is a mycelial endospore forming bacterium which is an obligate parasite of large number of nematodes with a life cycle remarkably synchronizing with the development of its host. Infection occurs through direct penetration of the cuticle by germinating spores that encumber to the surface of the nematodes. The life cycle begins with the vegetative mycelial colonies which eventually fill the nematode body with a large mass of bacterial endospores. Spores are liberated in the soil from the body of the nematode upon decomposition. An infected female can harbour up to 2 million endospores.

The resistance of the long lived spores to heat and desiccation and their favourable compatibility with nematicides, are characteristics well suited for their eventual practical use in field soil. However, the obligate nature of the organism, its lack of mobility and its dependence on dispersal agents are its major drawbacks. *Pasteuria penetrans* has been demonstrated to exert effective biocontrol of root-knot nematode on tomato. There has been a significant reduction in the root gall index and the number of egg masses of *Meloidogyne incognita* by about 52 and 60 percent respectively due to *P. penetrans* antagonism in brinjal. *P. penetrans* offers potential for the biocontrol of *Tylenchulus semipenetrans* infecting citrus. Black pepper plants inoculated with *P. penetrans* showed reduction in *Meloidogyne incognita* population and root gall indices and improved plant growth parameters.

A number of strains are reported to exist in *Pasteuria penetrans.* An isolate of the parasite infecting *M. incognita* did not encumber *Pratylenchus brachyurus* and two isolates that adhered to *Pratylenchus brachyurus* could not adhere to *M. incognita.* In India, two different strains of *P. penetrans* have been reported *i.e.* one infecting *M. incognita* but not *Heterodera cajani* and the other infecting *H. cajani* but not *M. incognita. P. penetrans* reduces root penetration by the nematode juveniles, inhibits egg formation, reduces nematode penetration in soil and results in increase in yield. The bacterium being resistant to heat, desiccation, its favourable compatibility with nematicides and its ability to survive for more than 2 years in soil, qualifies it as the most potential bacterial biocontrol candidate which could be used on a large scale under field condition.

Fungi

Interactions between fungal antagonists and nematodes have been known to occur in agricultural soils for many years. Nematode destroying fungi play an important role in regulating nematode population dynamics. Recent studies have indicated the existence of opportunistic soil fungi which are capable of colonizing females and eggs of nematodes and deleteriously affecting them. Nematodes belonging to such genera as *Meloidogyne, Tylenchulus* and *Rotylenchulus,* at sedentary stages of their life cycle, are vulnerable to attack by fungi either within host plant roots or when females and egg masses are exposed on the root surface or within soil. Fungal growth is known to be enhanced in the rhizosphere, increasing the likelihood of fungal invasion of nematode structures present in the ecological niche. Once in contact with eggs, larvae or females, the fungi grow rapidly and eventually parasitize all the eggs that are in the early embryonic developmental stages.

Several fungi such as *Paecilomyces lilacinus, Verticillium lecanii, Pochonia chlamydosporia, Trichoderma harzianum, T. viride,* and *Gliocladium virens* have been reported to parasitize eggs of plant parasitic nematodes (Fig. 17). *Paecilomyces lilacinus* is a nematophagous fungus which is a wide spread facultative parasite of the sedentary stages of plant parasitic nematodes. The fungus colonizes the root surface and parasitizes eggs and egg-masses of root-knot nematodes. Semi-selective medium for the isolation of this bio-agent from soil and roots have also been standardized.

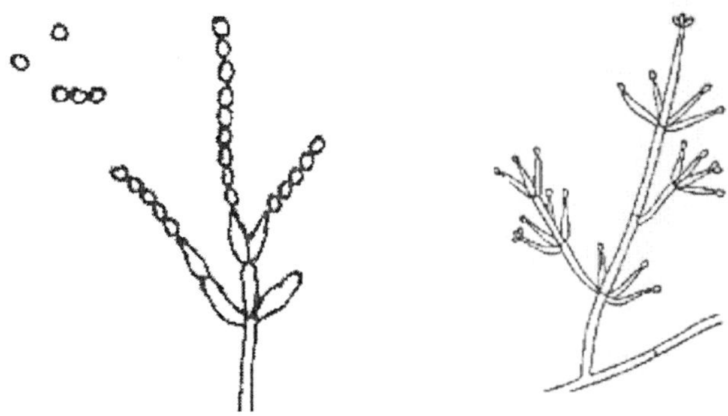

a b

Fig. 17 : Some fungi used for the biocontrol of plant parasitic nematodes a. *Paecilomyces lilacinus*, b. *Pochonia chlamydosporia*

Paecilomyces lilacinus

Paecilomyces lilacinus is by far the most promising and practicable biocontrol agent for the management of root-knot and cyst nematodes (Plate 18a). The fungus grows abundantly in sterilized paddy seeds, mashed potato, chopped water lily plants, ipil ipil (*Leucaena glauca*) leaves and corn grits. Eggs of *Meloidogyne incognita* on potato roots were found to be infected by *P. lilacinus*. This fungus was also capable of invading mature female and cysts of *Globodera pallida*. *Paecilomyces lilacinus* increased growth of brinjal plants and reduced root galling, egg masses and number of eggs per egg mass of *M. incognita*. Hyphae of *P. lilacinus* readily penetrate egg shells of *M. arenaria* through small pores dissolved in the vitelline layer. The effect seems to involve disruption of embryonic development and death of larvae, resulting in reduction of nematode populations.

Paecilomyces lilacinus was most effective in the management of root-knot nematodes in black pepper vines. *P. lilacinus* suppressed *Meloidogyne* spp. population by 58.3 - 86.9% in cardamom nursery. The number of quality seedlings for transplanting was greater in beds treated with this biocontrol agent. *P. lilacinus* at 4 and 8 g / kg soil significantly reduced *M. incognita* population in soil and root-knot index and increased number of leaves per plant, weight of 100 leaves, shoot and root weight of betelvine. *P. lilacinus* was also effective for the management of reniform nematode infecting tomato and brinjal.

Pochonia chlamydosporia

Various investigators have reported *Pochonia chlamydosporia* (= *Verticillium chlamydosporium*) as a potential biocontrol fungus against root-knot nematodes (Plate 18b). Methods have been already standardized for studying the growth and survival of this fungus. This biocontrol fungus was found to parasitize the eggs and females of root-knot nematodes. *P. chlamydosporia* is capable of preventing egg hatching of root-knot nematodes by its hyphal penetration. The bio-agent was found to be effective in the control of root-knot nematodes on various crops. To achieve the desirable control of the root-knot nematodes, the colonization of root-system by *P. chlamydosporia* was found to be very important and this colonized bio-agent infects the eggs produced by first generation and second generation nematodes. The fungus has also been found in the cysts of *Heterodera glycines*.

Trichoderma spp.

These fungi suppressed *Meloidogyne* spp. by 58.3-68.9% in cardamom nursery and increased the number of quality seedlings for transplanting. The main types of destructive activity of opportunistic fungi are thought to be enzymatic disruption of nematode egg shells and larval cuticles and physiological disturbances brought about by biosynthesis of diffusible toxic metabolites.

a b

Plate 18 : Fungi for the biocontrol of plant parasitic nematodes a. *Paecilomyces lilacinus*, b. *Pochonia chlamydosporia*

The feasibility of using selected opportunistic soil fungi such as *Paecilomyces lilacinus, Pochonia chlamydosporia, Verticillium lecanii, Trichoderma harzianum, T. viride* and *Gliocladium virens* have shown some effect in controlling *Meloidogyne incognita, Tylenchulus semipenetrans* and *Rotylenchulus reniformis.*

Arbuscular mycorrhizal fungi

Arbuscular mycorrhizal (AM) fungi are geographically ubiquitous and occur over a broad ecological range. They are commonly found in association with agricultural crops, most shrubs, most tropical tree species and some temperate tree species. AM fungi have been observed in 1,000 genera of plants representing some 200 families. There are atleast 300,000 receptive hosts in the world flora and around 200 species of AM fungi. These fungi are obligate biotrophs. Increased plant growth because of AM fungal colonization is well documented. The increased plant growth is attributed to enhanced uptake of diffusion limited nutrients, hormone production, biological nitrogen fixation, drought resistance and suppression of root pathogens. Several workers have reported that AM fungi can act as biocontrol agents for alleviating the severity of disease caused by root pathogenic fungi, bacteria and nematodes.

Diversity of interactions between AM fungi and plant parasitic nematodes has been observed, but in most cases mycorrhizae have reduced the severity of disease caused by plant parasitic nematodes. Most AM fungi and nematode interaction studies have been made with sedentary endoparasites, especially with the root knot nematode belonging to the genus *Meloidogyne*. Some studies have shown that colonization of roots by AM fungi do not affect the penetration of roots by nematode juveniles of sedentary endoparasite *Meloidogyne* spp. in tomato and white clover, while there are a few reports indicating that AM colonized roots reduced the penetration of tomato, cotton, white clover roots by juveniles of *Meloidogyne* spp. Reduced penetration of AM colonized roots by the migratory endoparasites, *Pratylenchus brachyurus* (soyabean cyst nematode) and *Radopholus similis* (banana burrowing nematode) has also been reported. The development of juveniles to mature females ovipositing eggs was delayed and the number of eggs formed per egg sac was reduced in AM colonized roots. Reduced reproduction evidenced by suppressed egg and cyst production, reduced number of nematodes per gram of roots, number of galls per plant in case of AM colonization of roots have been reported by several workers.

The timing of AM fungal and nematode inoculation may influence the degree to which AM fungi protect plants against nematodes. Most of the studies conducted on staggered inoculation of AM fungi and plant parasitic nematodes brought out that application of AM fungi two weeks earlier than the nematode were better than simultaneous application of AM fungi and the nematode in reducing the severity of disease. Application of nematode two weeks earlier than the AM fungi was not effective in controlling the disease. AM fungi applied in the form of root inoculum are known to colonize the roots much faster than those applied as spores, since spores have to germinate and then colonize roots.

Mechanisms of suppression of nematodes by AM fungi

Studies conducted so far suggest that the mechanisms of suppression may be due to morphological, physiological and biological alterations in the host.

Morphological alterations

Thickening of cell walls through lignification in mycorrhizal plants thereby preventing penetration of tomato roots by *M.incognita* has been observed. Histopathological studies of nematode galls caused by *Meloidogyne* spp. showed that galls in mycorrhizal plants had fewer giant cells or syncitia, which are needed for the development of nematode larvae, compared to non-mycorrhizal plants. The nematodes in mycorrhizal plants were smaller and took a longer time to develop into adults.

Physiological and biochemical alterations

Decreased root exudation in mycorrhizal plants, because of increased membrane phospholipids content, possibly helps in reducing the infection by root pathogens. Increased phenylalanine and serine in tomato roots due to inoculation with *G.fasciculatum* was found to be inhibitory to root knot nematodes. The activation of specific plant defense mechanisms as a response to AM colonization is an obvious basis for the protective capacity of AM fungi. Among the compounds involved in plant defense studies in relationship to AM formation are phytoalexins, enzymes of phenylpropanoid pathway, chitinases, β-1, 3- glucanases, peroxidases, pathogenesis related (PR) proteins, callose, hydroxyproline-rich glycoproteins (HRGP) and phenolics. These plant defense mechanisms are activated to a very little extent by AM colonization. However these compounds could sensitize the roots to pathogens and enhance mechanisms of defense to subsequent pathogen infection.

Biological alterations

Mycorrhizal plants harbour higher population of microorganisms in the rhizosphere thus making it difficult for the pathogen to compete and gain access to the root. AM fungi could affect nematode activity indirectly through their effect on other rhizosphere populations. The AM fungi could enhance the parasitic activity of microorganisms such as, *Pasteuria penetrans* and *Pochonia chlamydosporia* against plant parasitic nematodes.

Like most instances of biological control AM fungi cannot offer complete immunity against the infestation by plant parasitic nematodes. They could only impart a degree of resistance against soil-borne parasitic nematodes. However, the possibility of biologically controlling plant parasitic nematodes looks promising.

References

Aggarwal, A. and Mehrotra, R.S. 1997. Biological control of plant pathogens. In: *New Approches in Microbial Ecology*. (eds. J.P. Tewari, G. Saxena, N. Mittal, I. Tewari, and B.P. Chamola). Aditya Books Private Ltd., New Delhi, India. pp.412-426.

Alexander, V.S. and Slevin, J. 2003. *Agrobacterium tumefaciens* mediated gene transfer: An overview. In: *Biotechnology, Emerging Trends* (eds. J.Selvin, A.S.Ninawe, V.S.Sugunan, N.Sukumaran and A.P. Lipton). Biotech Books, New Delhi, India. pp.93 101.

Altman, A. (ed.) 1998. *Agricultural Biotechnology*. Marcel Dekker, New York, USA.

Bagyaraj, D.J. 2003. Mycorrhizal Biotechnology for improved crop productivity. In: *Biotechnological strategies in Agro-Processing*. (eds. S.S. Marwaha, and J.K. Arora). Asiatech Publishers, New Delhi, India. pp. 80-84.

Bagyaraj, D.J. and Govindan, M. 1996. Microbial control of fungal root pathogens. In: *Advances in Botany*. (eds. K.G. Mukerji, B. Mattur, B.P. Chamola, and P. Chitralekha). APH Publishing Corporation, New Delhi. pp. 293-322.

Brodie, I.D.S. and Blakeman, J.P. 1975. Competition for carbon compounds by a leaf surface bacterium and conidia of *Botrytis Cinerea*. *Physiological Plant Pathology*. 6: 125-135.

Chahal, P.P.K. and Chahal, V.P.S. 2004. Biocontrol agents for the management of nematodes. In: *Plant Diseases - Biocontrol Management*. (ed. Sampat Nehra) Aavishkar Publishers, Jaipur, India. pp. 295-307.

Gupta, R.P., Kalia, A. and Kapoor, S. 2007. Bioinoculants - A step towards sustainable Agriculture, NIPA Publishing Agency, New Delhi, India.

Helbig, J. 2006. Microbial control of plant pathogens of horticultural crops - An overview.In:*Microbial Biotechnology in Horticulture*. (eds. R.C. Ray, and O.P. Ward) Science Publishers, Enfield, USA. pp. 83-123.

Khan, M.W. 1998. Biopesticides in the management of plant parasitic nematodes. In: *Trends in Microbial Exploitation*. (eds. B. Rai, R.S. Upadhyay and N.K. Dubey). International Society for Conservation of Natural Resources, Varanasi, India. pp. 139-149.

Li, B., Xie, G. and Soad, A.C. 2005. Suppression of *Meloidogyne javanica* by antagonistic and plant growth-promoting rhizobacteria. *Journal of Zhejiang University, SCI*. 6B: 496-501.

Mukerji, K.G. and Ciancio, A. 2007. Mycorrhizae in the integrated pest and disease management. In: *General concepts in Integrated Pest and Disease Management* (eds. A. Ciancio, and K.G. Mukerji). Springer, Dordrecht, The Netherlands. pp 245-268.

Nakas, J.P. and Hagedorn, C. (eds.) 1990. *Biotechnology of Plant Microbe Interactions.* McGraw-Hill Publishing Co., New York, USA.

Pal, K.K., Tilak, K.V.B.R., Saxena, A.K., Dey, R. and Singh, C.S. 2000. Antifungal characteristics of a fluorescent *Pseudomonas* strain involved in the biological control of *Rhizoctonia solani. Microbiological Research.* 155: 233-242.

Pal, K.K., Tilak, K.V.B.R., Saxena, A.K., Dey, R. and Singh, C.S. 2000. Monitoring of a biocontrol rhizobacterium *Pseudomonas glumae* EM85 using Tn5::*lacZ* marker *Indian Journal of Microbiology.* 40: 21-24.

Prasad, D. 2004. Progress and problems in the use of *Pasteuria penetrans* and VAM for the management of plant parasitic nematodes. In: *Plant Disease – Biocontrol Management.* (ed. Sampat Nehra) Aavishkar Publishers, Jaipur, India. pp. 285-294.

Ramanujam, B. and Rabindra, R.J. 2006. *Current Status of Biological Control of Plant Diseases Using Antagonistic Organisms in India.* Project Directorate of Biological Control Pub., Bangalore, India.

Sampat Nehra. 2004. Integrated management of plant parasitic nematodes. In: *Plant Diseases - Biocontrol Management.* (ed. Sampat Nehra) Aavishkar Publishers, Jaipur, India. pp. 367-399.

Saxena, G. and Mittal, N. 1997. Ecology of nematode-destroying fungi In: *New Approaches in Microbial Ecology.* (eds. J.P. Tewari, G. Saxena, N. Mittal, I. Tewari and B.P. Chamola). Aditya Books Private Ltd., New Delhi, India. pp. 39-86.

Sheela, M.S., Nisha, M.S. and Manu, C.R. 2004. Fungi - A tool for nematode management. In: *Plant Diseases - Biocontrol Management.* (ed. Sampat Nehra) Aavishkar Publishers, Jaipur, India. pp. 185-207.

Siddiqui, Z.A., Mahmood, I. and Khan, M.W. 1999. VAM fungi as prospective biocontrol agents for plant parasitic nematodes. In: *Modern Approaches and Innovations in soil Management,* (eds. D.J. Bagyaraj, K.K. Khanna, A. Varma, and H.K. Kehri). Rastogi publications, Meerut, India. pp. 47-58.

Trivedi, P.C. 2002. Biological control of plant parasitic nematodes: Current status. In: *Glimpses in Plant Sciences* (eds. K.R. Aneja, M.O. Charaya, A. Aggarwal, D.K. Hans and S.A. Khan). Pragati Prakasthan, Meerut, India. pp. 146-152.

Vallad, G.E. and Goodman, R.M. 2004. Systemic acquired resistance and induced systemic resistance in conventional agriculture: review and interpretation. *CropScience.* 44: 1920-1934.

Whipps, J.M. 2001. Microbial interactions and biocontrol in the rhizosphere. *Journal of Experimental Botany.* 52: 487-511.

Chapter-7

Microorganisms for Biocontrol of Insect Pests

The adoption of modern agricultural technologies has culminated into serious problems of pesticide resistance in major pests, resurgence in minor pests, elimination of populations of natural parasites, predators and pollinators, environmental pollution, pesticide residues in food chain and health hazards. World Health Organization (WHO) estimates one million pesticides poisoning cases and 20,000 deaths every year globally. This is due to high pesticide residues in food chain including cereals, pulses, vegetables, fruits, milk and milk products (including mother's milk), fishes, poultry, meat products and water. Twenty five per cent of Indian food products contain pesticide residues above tolerance level compared to only 1.2% globally. Moreover being more resistant to degradation by abiotic and biotic agencies the pesticide residues are observed to leach down to lower strata of soil, easily get absorbed by plant roots, accumulate in the food chain to be biomagnified in the food web. This feature of biomagnification of pesticides in plant and animal tissues (particularly lipid bodies) makes their use to be health hazardous as this may lead to several ailments. Chemical pesticides cause significant health hazards to man including convulsions, diarrhoea, dermatitis, allergies, anemia, organ dysfunctions, blindness, immunosuppression, neurotoxicity, vasodilation, impairing of reproductive functions, leukaemia, carcinogenecity, tumorogenecity and paralysis.Apart from these, they also cause harm to non target beneficials, soil microflora and crops.

Most of the chemical pesticides presently used fall under extremely hazardous and highly hazardous categories. In our current climate of increased awareness of the sensitivity of our environment, among both the scientific community and the general public, the development of environmentally safer methods of insect control is desirable. Tightened registration procedures around the world will likely result in a decreased availability of a number of broad-spectrum pesticides, creating new opportunities for alternative methods (including microbials). An integrated pest management (IPM) strategy, in which natural enemies of pest arthropods and other alternative measures play significant roles in crop protection, will minimize negative environmental impacts and other deleterious effects due to insecticide usage while providing a more sustainable approach to pest control. Several entomopathogens (viruses, bacteria, fungi, and nematodes) offer effective means of microbial control that can be combined with other tactics such as mating disruption and the use of reduced-risk pesticides. In addition, microbial control agents (MCAs) are safe for the environment, beneficial insects, applicators, and the food supply, and they can be applied just prior to harvest.

Historical Background

History reveals that Aristotle in 320 B.C. was the first to mention about an insect disease in his book 'Historia animalium'. Bassi in 1834 demonstrated for the first time that a microorganism, the fungus *Beauveria bassiana* could cause an infectious disease in animal (silkworm). He also suggested that such infectious organisms could be used to control insects. Metchnihoff from Russia in 1879 for the first time mass produced the insect pathogenic fungus *Metarhizium anisopliae* and used it as a microbial insecticide against grain weevil and sugar beet curlio. Intentional control of insect pests with bacteria was first attempted by d'Herelle in 1914. The use of viruses to control insect pests was stimulated by the efforts of Balch *et al.* and Steinhaus following world war II. Another major contribution to the development of microbial insecticides was the establishment of an Insect pathology laboratory at the University of California in 1945 by Dr. E.A. Steinhaus. Both successes and failures were reported in the beginning of twentieth century and the inconsistency of the results retarded the usage and development of microbial insecticides. The situation has changed today. Intensive evaluation of microbial insecticides is taking place at present in many countries through out the world. The interest in microbial insecticides has expanded because of problems such as insect resistance, emergency of secondary pests, and toxic residues that have developed with the use of chemical insecticides. Moreover, recent successes with microbial insecticides have greatly encouraged their usage.

Prerequisites of a Microbial Insecticide

A microbial insecticide should possess certain basic prerequisites before it can be considered for practical field use.

(i) It should be highly virulent for the target insect.

(ii) It should be stable to stresses of nature.

(iii) Its virulence should be confined to the species or group of insects which is being treated.

(iv) The lethality should be rapid so that the crop damage can be held to a minimum.

(v) It should neither damage the crops nor should it be harmful to the beneficial insects, predators and higher animals.

(vi) The organism or its product should be amenable to easy production in conventional industrial equipment by processes which are capable of yielding an economic product.

(vii) The product should be stable over an extended period so that it can be stored and transported to its site of use.

Because of these restrictive qualities assigned to these biological control agents, the number of microbial insecticides produced on a large scale are few. Organisms commonly used at present as microbial insecticides include bacteria like *Bacillus thuringiensis, Bacillus popilliae, Pseudomonas fluorescens*, fungi like *Beauveria bassiana, Metarhizium anisopliae,* viruses like Nucleopolyhedrosis viruses (NPV), Granuloviruses (GV), protozoa like *Nosemia fumiferarae* and nematodes like *Steinernema* sp., *Heterorhabditis* sp. etc. The advantage of using microbial insecticides over chemical pesticides are due to their eco-safety, target specificity, non-development of resistance, reduced number of applications, yield and quality improvement, higher acceptability and value of produce for exports and suitability for rural areas.

Biopesticides are products based on living biologicals which are prone to decay/degradation in soil and water after 20-30 days. The chemical pesticides, however, persist in air, soil and water for very prolonged periods causing environmental pollution and health hazards. Biopesticides kill the target pests only and do not harm other insects. For example, *Bacillus thuringiensis* var. *kurstaki* kills only lepidopteran caterpillars and has no impact on pollinators, honeybees or adult moths. The injudicious and indiscriminate use of chemical pesticides has led to the development of resistance in the pests and therefore, the farmers have to resort to repeated applications of variety of pesticides. Contrary to this, only 2-4 applications of biopesticides may be required as they are highly efficacious with no development of resistance. Another advantage of biopesticides is that they do not cause any changes in the biology of minor pests, whereas the chemical pesticides are known to have resulted in the resurgence of many minor pests of yester years.

Helicoverpa armigera was initially known as "gram pod borer", but, currently it causes damage to almost all crops. The use of biopesticides results in economic gains such as increased yields, improvement of quality, extra shine, better taste and higher price for the produce. The exportable agriculture products like cotton, fruits, vegetables, tobacco, tea etc. with certification of origin from biopesticide/organic farming areas find enhanced acceptability and fetch higher price of the produce. The biopesticides being eco-safe and harmless to human beings are the most ideal for use in rural areas where the labour/marginal farmers are not only ignorant about the precautions to be taken during chemical pesticide application operations, but, at times they do not have the capability of purchasing such protective gears.

Fungi for the Management of Insect Pests

A diverse spectrum of fungi is reported from insects and mites. Some species in the Entomophthorales (e.g., *Neozygites fresenii, Entomophaga maimaiga*) and Hypocreales (e.g., *Lecanicillium* spp., *Aschersonia* spp., and *Hirsutella* spp.) produe epizootics in pest populations that often lower pest densities to levels that are not harmful to the crop. Commercial development of entomopathogenic fungi has been confined to species in the Hypocreales, most notably *Lecanicillium* spp., *Beauveria bassiana, Metarhizium anisopliae,* and *Paecilomyces fumosoroseus.* In general, these fungi are inundatively applied to bring about a rapid reduction in the pest population. Because fungi gain access to the host through the cuticle, they are the principal MCAs of sucking insects (Hemiptera, e.g., aphids and whiteflies) and control a variety of insects in other orders such as Coleoptera, Lepidoptera and Orthoptera. Fungi commonly used for the biocontrol of insect pests are given in table 21.

Table 21: Entomopathogenic fungi used against different pests

Entomopathogens	**Insect-pests**
Metarhizium anisopliae	Sugarcane spittle bug; sugarcane soldier fly; scarabaeiid grub
Metarhizium flavoviride	Brown plant hopper
Beauveria bassiana	Pine caterpillar; black vine weevil; citrus root weevils; corn borer; eriophid mites; clear winged grasshopper; beetle; fire-ant
Beauveria brongnartii	European cock chaffer
Entomophaga grylii	White fly
Pandora neoaphidis	Pea aphid
Entomophthora muscae	Muscoid flies; onion fly
Zoophthora sp.	Clover weevil
Zoophthora phytonomi	Clover weevil
Zoophthora radicans	Spotted alfalfa aphid
Coelomomyces stregomyiae	Mosquito
Coelomomyces iliensis	Mosquito
Entomophaga anlicae	Brown tail moth
Entomophaga maimaige	Gypsy moth larvae
Entomophthora erupta	Green apple bug
Hirsutella thompsonii	Citrus rust mite
Hirsutella citriformis	Brown plant hopper
Paecilomyces fumosa-rosea	Sweet potato white fly
Nomuraea rileyi	Noctuid defoliators
Paecilomyces lilacinus	Brown plant hopper
Aschersonia aleyrodis	White fly larvae
Verticillium lecanii	Aphids, whitefly
Lagenedium giganteum	Mosquito
Culicinomyces clavisporus	Mosquito

Conidium is the infective unit in entomogenous fungi, which may penetrate the insect cuticle from a combination of mechanical pressure by the germ tube and enzymatic degradation of the cuticle. Strains of these fungi produce proteases and chitinases in liquid culture. It is interesting that proteases are produced first in cultures, which is consistent with the requirement for degradation of the protein matrix before the fungus can degrade chitin. Once through the cuticle, and having overcome host defense mechanisms, the fungus proliferates as hyphal bodies, which multiply by budding. The fungus spreads through the haemocoel and the insect normally dies 3-14 days after spore application. Toxins produced by these fungi also play a significant role in their pathogenicities and a number of such allelochemicals have been isolated and identified.

Metarhizium anisopliae

This hyphomycetous fungus is commonly called as the green-muscardine fungus. In 1879, Metchnikoff isolated the fungus from the beetle *Anisopliae austriaca* and suggested its use as a microbial agent against insect pests. The colony of *Metarhizium anisopliae* appears white when young, but as the conidia mature, the colour turns to dark green (Plate 19a). The conidiophores are branched, and the initial conidium is produced by simple abstriction at the distal end of the conidiophore (Fig. 18 a). A chain of conidia is formed on each conidiophore with the youngest conidium being adjacent to the conidiophore. The mass of spore chains becomes so dense and coheres with each other to produce prismatic masses of columns of spore chains. *Metarhizium anisopliae* infects a wide range of insects including members of Coleoptera, Lepidoptera, Diptera, Homoptera and Collembola (Plate 19b, c).

It was the first fungus to be produced in a large scale from a biological control perspective. *M. anisopliae* produces toxic proteolytic enzymes and esterases and also a number of secondary metabolites which are mycotoxins such as destruxins, cyclodepsipeptides and desmethyldestruxin B. Destruxins initially cause paralysis and finally death of the insect. The fungus is mass produced commercially and used for the biocontrol of insect pests like sugarcane spittle bug, *Mahanarva postica*; coconut rhinocerous beetle, *Oryctes rhinoceros*; diamond back moth on cauliflower, *Plutella maculipennis*; *Pyrilla* infestation in sugarcane etc.

a

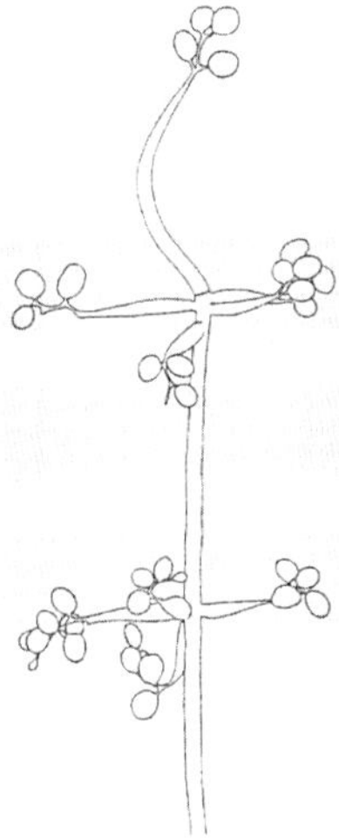

b

Fig. 18 : Some fungi for biocontrol of insect pests a. *Metarhizium anisopliae,* b. *Beauveria bassiana*

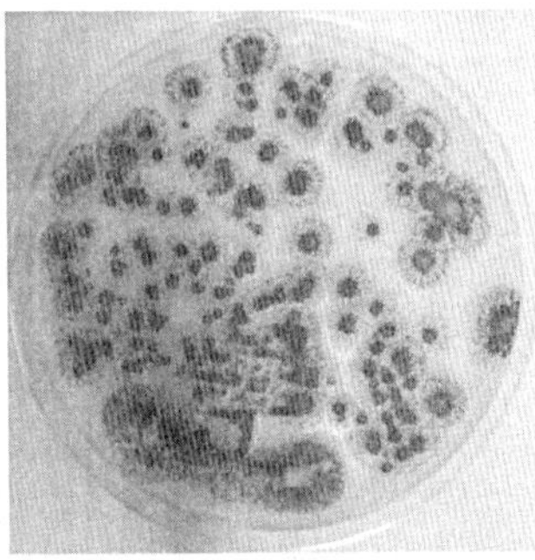

a

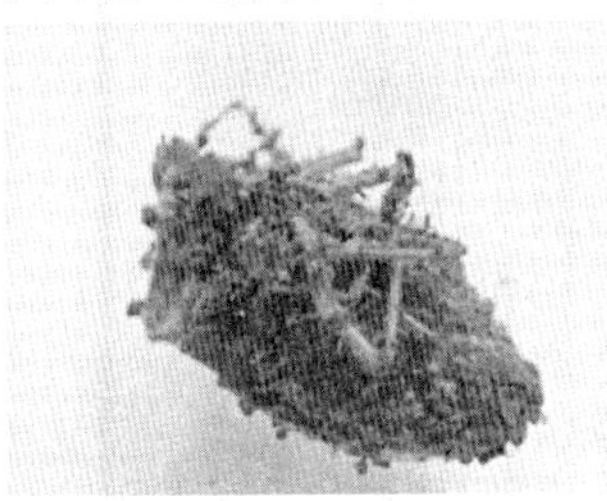

b c

Plate 19 : Insect pathogenic fungus *Metarhizium anisopliae* a. Plate showing growth, b,c. Diseased insects

Beauveria bassiana

Beauveria bassiana, another hyphomycetous fungus causes white muscardine disease in insects. The white muscardine was the first disease in animals (silkworms) shown to be caused by a microorganism. The fungus produces a flat, chalky, pulverulent growth, with conidia formation taking place in 3-7 days (Plate 20a). The conidia are hyaline, one-celled, smooth, thin-walled, globose to oval, borne on sympodial conidiophores with a zig-zag appearance (Fig. 18 b). *B. bassiana* occurs world wide and has been recorded from more than 500 insect hosts. It also occurs in soil as ubiquitous saprophyte. The hosts are mainly Lepidoptera, Coleoptera, and Hemiptera, but also occur in Diptera and Hymenoptera. Some of the major economic insect pests that are susceptible to this fungus are the European corn borer, *Ostrinia nubilalis;* the codling moth, *Laspeyresia pomonella;* the Japanese beetle, *Popillia japonica;* the Colorado potato beetle, *Leptinotarsa decemlineata;* the chinch bug, *Blissus leucopterus;* and the European cabbageworm, *Pieris brassicae* (plate 20b).

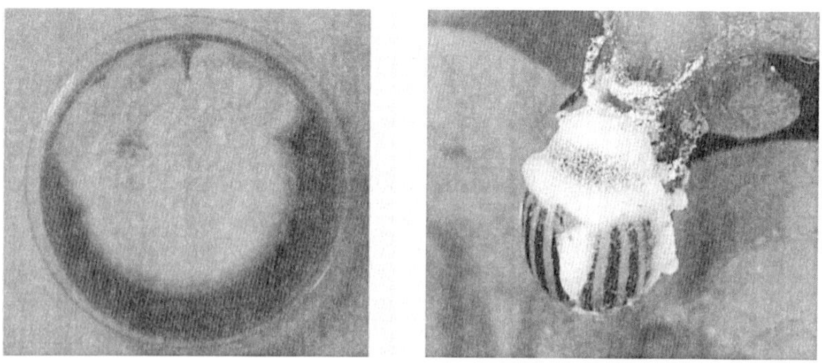

a b

Plate 20 : Insect pathogenic fungus *Beaveria bassiana* a. Plate showing growth, b. Diseased insect

The germination of the conidia of *B. bassiana* on the integument and in culture medium is dependent on certain nutritional requirements. The germinating conidia require an utilizable source of carbon, such as glucose, glucosamine, chitin, and starch, and also a nitrogen source is needed for hyphal growth. Infection generally takes place through the integument. In the fire ant, *Solenopsis richteri,* however, 37% of the conidia fed to larvae germinated in the digestive tract within 72 hours, and the hyphae penetrated the gut wall between

60 and 72 hours. The injury to the digestive tract enabled the digestive juices to enter the haemocoel and change the haemolymph pH. The white-muscardine fungus also invades preferentially the larval digestive tract of the corn earworm, *Helicoverpa zea,* to cause starvation and nutrient depletion that may be involved in larval death. *B. bassiana* produces the mycotoxins beauvericin and bassianolide and the composition of the media influences the production of mycotoxins. The fungus also produces certain secondary metabolites, which prevent bacterial putrefaction and thus permit mummification of the insect.

The fungus is commercially produced and available in the market for application in the field by farmers especially for the control of Colorado beetle, *Leptinotarsa decemlineata*; European corn borer, *Ostrinia nubilalis*; the pine caterpillar, *Dendrolimus punctatus*; sugarcane stalk borer, *Chilo auricilius*; sugarcane black bug, *Cavelerius sweeti*; the white grub, *Holotrichia serrata* etc.

Nomuraea rileyi

This hyphomycetous fungus is well known for its epizootics among Lepidoptera, though they were reported from Homoptera, Diptera and spiders. The diagnostic characters of the fungus include conidiophores mononematous to synnematous with vertically arranged cylindrical to oval short conidiogenous cells bearing conidia. Conidia are one-celled, smooth, ovoid to cylindrical, in short divergent chains, pale to dark green in colour.

N. rileyi has the ability to cause spectacular epizootics under favourable environmental conditions. *N. rileyi* epizootics has been observed on *Spodoptera exigua, Helicoverpa armigera, Junonia orthiya, Anticarsia gemmatalis, Spodoptera litura, H. zea, H. virescens* and *Trichoplusia ni*. On the insect initially the fungus appears white, later turning pale green or purple. Exposure of the infected larvae to light enhances conidial production. The fungus also produces a mycotoxin that is toxic to larva of the gypsy moth, *Lymantria dispar*. These epizootics generally result in a relatively heavy load of soil-borne conidia that act as natural reservoir for the next crop. Persistence of *N. rileyi* is reported to increase with the addition of sunflower and groundnut oils. Field trials in India showed that both foliar spray and soil application is very effective against *Spodoptera litura.*

Hirsutella thompsonii

The genus *Synnematium* is a synonym of *Hirsutella* and an imperfect stage of *Cordyceps*. This hyphomycetous fungus, on the insect cadaver produces sparse mycelium, thus may be overlooked by inexperienced workers. Conidiophores are synnematous bearing conidiogenous cells that are swollen at the base with slender necks. Conidia are one celled, ellipsoidal to cylindrical covered with persistent slime. *H. thompsonii* attacks only mites and is well known for its epizootics among eriophyid mites, particularly citrus rust mite (*Phyllocoptruta oleivora*), citrus bud mite (*Eriophyes sheldonii*) and coconut mite (*Eriophyes guerreronis*). It is known to have considerable potential as an acaricide and is commercially produced.

Bacteria for the Management of Insect Pests

Various spore-forming and non-spore forming bacteria are pathogens of insects, but only *Bacillus thuringiensis* has been studied extensively. *B. thuringiensis* has been employed to control a multitude of insects in organic and conventional agriculture, horticulture, forestry, and public health. Its safety to pesticide applicators, the food supply, beneficial organisms, and the environment is thoroughly documented. A variety of *B. thuringiensis* formulations are commercially produced in several countries with insecticidal activity for species of Lepidoptera, Coleoptara and Diptera. The insecticidal activity of *B. thuringiensis* is due to the crystal protein toxins contained in the parasporal inclusions that are produced at the time of sporulation. These toxins basically function as stomach poisons and kill the insect by disrupting osmotic balance in the midgut epithelium. Damage to the midgut epithelium results in cessation of feeding and gut paralysis and ultimately death. *B. thuringiensis* products are made up primarily of spores and toxins, thus differing from most other MCA products (e.g., nematodes, fungi, and viruses), which contain the entire organisms. Bacteria commonly used for the control of insect pests are given in table 22.

Table 22 : Some bacterial pesticides currently used

Bacterial agents	Target organisms
Bacillus popilliae	Japanese beetle grub of lawn and pastures
Bacillus thuringiensis	Caterpillar pests of vegetables, forest trees and ornamentals
Bacillus thuringiensis sub sp. *israelensis*	Mosquito larvae and black flies
Bacillus thuringiensis sub sp. *kurstaki*	Lepidopteran larvae
Bacillus thuringiensis sub sp. *aizawai*	Wax moth caterpillar
Bacillus thuringiensis sub sp. *tenebrionis*	Beetles
Bacillus sphaericus	Mosquito larvae

Bacillus thuringiensis

Bacillus thuringiensis (Bt) is a Gram-positive, spore forming, entomopathogenic bacterium characterized by the production of proteinaceous crystalline inclusions during the stationary phase of its growth (Fig. 19; Plate 21a). Bt has been found to be an inhabitant of various ecological niches including soil, insect cadavers, dust of stored products and plant surfaces.

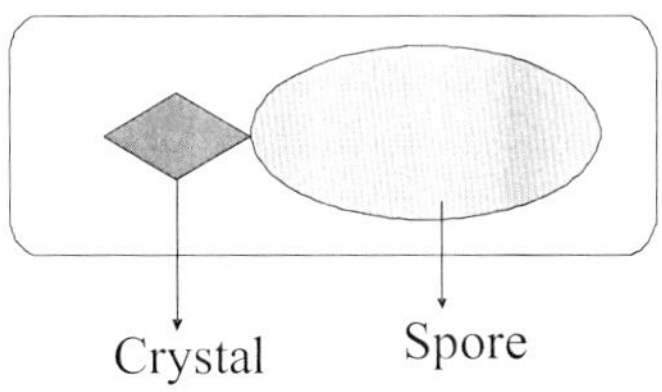

Fig. 19 : *Bacillus thuringiensis* cell showing spore and crystalline inclusion

The entomopathogenicity of Bt has been largely or completely attributed to the parasporal crystal inclusions, also known as insecticidal crystal proteins (ICPs), Cry proteins or as δ-endotoxins. The cry proteins that have been studied so far are toxic to various pests, including vectors of human diseases, but are also non-pathogenic to non-target insects, mammals, birds, amphibians and reptiles. Apart from δ-endotoxins, Bt strains also produce other enzymes and secondary metabolites such as phospholipases, heat liable β-exotoxins, proteases, chitinases and vegetative insecticidal proteins that contribute to their levels of virulence.

Although Bt was first observed by Ishiwata in Japan in 1902, its scientific description was recorded in 1911 by Berliner in Germany.

Steinhaus in USA established the commercial potential of Bt in 1950. Several strains of Bt have been identified worldwide from numerous screening programmes. Bt isolates have narrow specificities against various insects, but together they span a wide range of orders including Lepidoptera (beetles and moths), Diptera (mosquitoes and blackflies), Coleoptera (beetles and weevils), Hymenoptera (wasps and bees), with some isolates also active against nematodes, mites and protozoa. There exists 34 recognized subspecies of *B. thuringiensis* including most commonly used subspecies *kurstaki, israelensis,* and *tenebrionis*. The crystal proteins are biodegradable and are specific to each subspecies/ strain and are coded by a single plasmid borne gene. The crystal proteins could be classified on basis of amino acid sequence homology and presently fall under 40 different classes.

The primary action of Cry toxins is to lyse mid gut epithelial cells in the target insects by forming lytic pores in the apical microvillar membranes. Unlike most chemical insecticides that function on contact, Bt insecticides in the form of a spray or a Bt crop must be ingested by the target organism to be effective. The mechanism of action of Cry toxins can be broadly divided in to three steps: (i) solubilization and activation of protoxin, (ii) recepter recognition and binding and (iii) formation of lytic pores and cell lysis. Upon ingestion, the Cry protoxins are solubilized in the alkaline/acidic gut environment of the target insect, followed by its activation by specific proteases present in gut. The activated toxin then binds to specific receptors present on the midgut epithelium which further leads to the formation of lytic pores, cell lysis and consequently death of the host. Each step can modulate the activity of a Cry toxin against particular insect and hence the toxins overall specificity. Large differences in the gut physiology among insect orders can also be responsible for the variations in the proteolytic activity of target pests and, hence, in the different cry toxins affinities. A full list of all the *B. thuringiensis* toxins discovered so far is available at http://www.biols.susx.ac.uk/Home/Neil_Crickmore/Bt/ Index.html.

Bt preparations are effective against the following insect pests; common cabbageworm, diamond-back moth, fall webworm, oriental tobacco budworm, persimmon fruit moth, rice skipper, Japanese tussock moth, sweet potato leaf worm and leaf folder gypsy moth, corn earworm, orange dog butter-fly, green rice caterpillar, soybean pod borer, cabbage webworm, cherry hornworm, black riceworm, the pine processionary moth, the European pine short moth and several others (Plate 21b). Bt preparations are available as wettable powder dusts and water-dispersible emulsions. Till date, Bt based formulations (67 registered products with more than 450 formulations) occupy key position accounting for nearly 90% of total biopesticide sales. *B. thuringiensis* based products are cost effective, very specific to target pest species, and safe to people and environment.

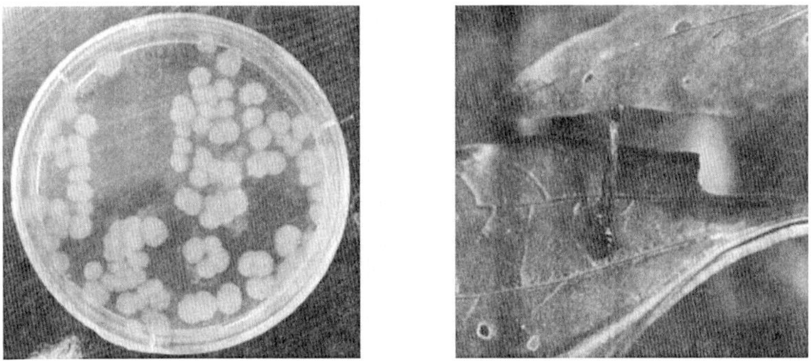

a b

Plate 21 : Insect pathogenic bacterium *Bacillus thuringiensis* a. Plate showing growth, b. Diseased insect

Bacillus popilliae

This bacterium is very effective in the management of the Japanese beetle, *Popillia japonica.* It causes milky disease, which is based on the milky white appearance assumed by the infected grubs. Grubs are susceptible to the pathogen, but not the adults. In addition to Japanese beetle other scarabaeid larvae like *Anomala orientalis, Autoserica castanea, Cyclocephala borealis* and *Strigodermella pygmaea* are also susceptible to the attack by *B. popilliae.* The bacterium is a spore-forming rod, facultative anaerobe with spindle-shaped spores. Infection of the grub occurs mainly through the ingestion of contaminated food. Ingested spores germinate, penetrate the gut wall and enter the haemocoel and multiply rapidly, sporulate and cause septicemia (bacteria in the haemolymph producing harmful effect on the host). Six days after infection the grub starts appearing milky white because of dense sporulation ranging from 2 to 5 billion spores per grub. The insect becomes sluggish and moribund. The grub eventually dies and serves as a source of inoculum in soil transmitting the disease. Since the bacterium does not produce spores when grown on artificial media, it is mass multiplied on the host, crushed and mixed with a carrier such as calcium carbonate to form spore dust. The spore dust is applied for the biological control of the Japanese beetle.

Viruses for the Management of Insect Pests

Several major groups of viruses infect insects and mites, but baculoviruses have received the most attention and commercial development. They are reported from Lepidoptera, Coleoptera, Diptera, and Hymenoptera. Two distinct groups, nucleopolyhedroviruses and granuloviruses, have been used for experimental and practical control of insect pests. The specificity of viruses used for control of lepidopteran pests is well known. The narrow host ranges of most entomopathogenic viruses can be both a limitation and advantage depending on the suite of arthropod pests in a particular ecosystem and on the controlling effects exerted by predators and parasitoids and other alternatives to broad-spectrum insecticides. Generally, sensitivity to solar degradation necessitates reapplication of virus preparations at fairly short intervals, particularly if the host is multivoltine and if pest population pressure is high. Shading and microhabitats on the host plant and in the environment can protect some viruses from solar degradation and enable long-term persistence. Entomopathogenic viruses are of special interest since they are relatively pest specific with little or no impact on the environment and are, therefore, ecologically attractive alternatives to chemical pesticides. One such group, the baculoviruses, has received the attention of many researchers and has been relatively well studied. These viruses contain a circular double stranded DNA inside a protein coat and include two groups the granuloviruses (GVs) and the nucleopolyhedroviruses (NPVs). GVs have one nucelocapsid per envelope and replicate either within the cytoplasm or inside the nucleus of the host cell.

NPVs occur as either single nucleocapsids embedded in a protein matrix (SNPV) or multiple nucleocapsids within a protein medium (MNPV) and in both cases they replicate exclusively within the nucleus. The proteinaceous coat referred to as polyhedrin gives protection to the viruses from environmental factors such as UV radiation and adverse weather conditions. The polyhedra dissolve in alkaline medium, characteristic of most lepidopteran species, releasing nucleocapsids. NPV infected larvae show delay in the molting thereby maintaining the larval stage for a longer period, which allows the virus to multiply many times. Most of the tissues of the insect such as the fat body, haemocytes and epidermis act as virus factories and the cells become completely filled with polyhedral inclusion bodies. Eventually, the cells lyse releasing the virus.

The liquefied host contaminates the foliage with high levels of NPV. When another larva feeds on the contaminated foliage it acquires the infection and the transmission continues. It has been observed that in some species such as the gypsy moth, *Lymantria dispar* infected larvae migrate to the apical parts of the trees and die. The most extensively studied baculovirus that is used for controlling lepidopteran pests is the alfalfa looper NPV (AcMNPV). This virus has served as the reference standard for the use of viruses in pest control. The genome of AcMNPV has been completely sequenced and expression systems using this virus has been developed and are being commercially marketed. These large, double- stranded, circular DNA viruses have a host range limited primarily to lepidopteran species. In addition, most NPV species infect a relatively few genera within Lepidoptera. Since NPVs are naturally occurring and have a narrow host range, they are being touted as one of the environmentally attractive biological control agents against many lepidopteran pests of agricultural and horticultural crops and forests.

NPV has been successfully used for the management of *Helicoverpa armigera* on tomato, chickpea, sunflower, cotton, groundnut and several other crops. The diseased insect usually hangs upside down and dies (Plate 22). It has also been used as a successful microbial insecticide for the control of soyabean caterpillar, *Anticarsia gemmatalis, Spodoptera litura, S. exigua. S. frugiperda* and pests of several agricultural and horticultural crops. In the forest ecosystem NPV effectively controlled gypsy moth, *Lymantria dispar*, the Douglas-fir tussock moth, *Orgyia pseudotsugata* and several other pests.

Plate 22 : Insect diseased by NPV

Granulosis virus

Granular viruses produce capsule shaped proteinaceous body called granule occluding a single virion. GVs are more specific in their host range than NPVs as they have been reported only from Lepidoptera.

The coding moth, *Cydia pomenella* is a worldwide key pest of apples, pears and walnuts. A granulosis virus isolated from *C. pomonella* was highly virulent to the pest and killed it rapidly, protecting fruit from economic damage in numerous field trials in several countries. Currently, commercial formulations of CpGV are available in France, Switzerland, Germany, Russia and several other European countries. GVs have been successfully used in the management of several pests like *Plutella xylostella* and *Pieris rapae* in cabbage, *Homona magnanima* in tea, *Erinnyis ello* in cassava and several other insect pests. Examples of commercially available viral pesticides for pest control are given in table 23.

Table 23 : Examples of commercially available viral pesticides for pest control

Host insect	Crop(s)
	Granulosis viruses
Adoxophyes orana	Apple
Adoxophyes sp.	Tea
Cydia pomonella	Apple, pears
Erinnyis ello	Cassava
Phthorimaea operculella	Potatoes
Pieris rapae	Cabbage
Plodia interpunctella	Stored almonds, raisins
Plutella xylostella	Cabbage
	Nuclear polyhedrosis viruses
Anagrapha falcifera	Cotton, vegetables
Anticarsia gemmatalis	Soybean
Helicoverpa armigera	Cotton, tomato
Helicoverpa zea	Cotton
Lymantria dispar	Forests
Orgyia pseudotsugata	Forests
Spodoptera exigua	Ornamentals, vegetables, garden pea, grapes
Spodoptera frugiperda	Maize
Spodoptera litura	Vegetables, cotton, rice, peanuts
Spodoptera sunia	Vegetables

A GV isolated from the potato tuber moth, *Phthorimaea operculella*, has shown high efficacy in protecting the potato crop in the field as well as potato tubers under storage. Current methods of pest control in stored grains and other products rely on fumigation and insecticide application. With the ban on the use of ethylene bromide and concerns with other chemicals, attention has been focused on baculoviruses and other microbial agents for the control of insect pests in stored products. The lepidopteran storage pests like *Ephestia cautella, Galleria mellonella* and *Ephestia kuehniella* are known to suffer from epizootic due to baculovirus infections. The granulosis virus of the Indian meal moth, *Plodia interpunctella* has the potential for long-term control in storage bins. A freeze-dried, powdered formulation of PiGV with a relatively long shelf life has been developed. This formulation could be applied as a spray or dust. The virus is now recommended for the control of *Plodia interpunctella* in stored almonds and raisins and has been patented by the USDA.

Production of the Baculoviruses through Cell Culture

More than 200 cell lines have been established from approximately 70 species of insects. The majority of these cell lines have been described from Lepidoptera, Diptera, Orthoptera, Hemiptera, Coleoptera and Hymenoptera. Many established cell lines from lepidopteran species have proved to be invaluable tool for the *in vitro* propagation of insect-pathogenic viruses. The NPVs for controlling *Helicoverpa armingera, H. zea, Orgyia leucostigma* and *Trichoplusia ni* are now propagated *in vitro* through cell culture. The GV used in the management of the insect pest *Cydia pomonella* has also been successfully produced in cell lines. If cell culture methods could be developed for the production of insect viruses, several advantages can be envisaged. Selected cell strains with desired production qualities and free from contaminating microorganisms could be stored as frozen materials to provide a stable supply of cells without the risk of genetic alteration. During production, the conditions for cell growth and virus replication could be controlled to assure a uniform final product.

Nematodes for the Management of Insect Pests

Entomopathogenic nematodes have been recovered from soils throughout the world and are very common soil organisms. Most research on biocontrol using nematodes has been on entomopathogenic nematodes (EPNs), which consist of the families Steinernematidae and Heterorhabditidae. EPNs kill their hosts with the aid of bacteria carried in the nematode intestine (steinernematids are associated with *Xenorhabdus* spp. and heterorhabditids are associated with *Photorhabdus* spp.).

Steinernematidae and Heterorhabditidae are obligate pathogens in nature. The only stage that survives outside of a host is the non-feeding third stage infective juveniles (IJs). The IJs carry cells of their bacterial symbiont in their intestinal track. After locating a suitable host, the IJs invade it through natural openings (mouth, spiracles, anus) or thin areas of the host's cuticle (common only in Heterorhabditidae) and penetrate into host haemocoel. The IJs recover from their developmental arrestment, release the symbionts, and bacteria and nematodes cooperate to overcome the host's immune response. The bacteria propagate and produce substances that rapidly kill the host and protect the cadaver from colonization by other microorganisms. The nematodes start developing, feed on the bacteria and host tissues metabolized by the bacteria, and go through 1-3 generations. Depleting food resources in the host cadaver lead to the development of a new generation of IJs that emerges from the host cadaver in search of a new host. A major difference between Steinernematidae and Heterorhabditidae is that *Heterorhabditis* adults are hermaphrodites in the first generation but amphimictic in following generations whereas *Steinernema* adults are always amphimictic.

Each nematode species is specifically associated with one symbiont species, although a symbiont species may be associated with more than one nematode species (Table 24). Many of the well studied entomopathogenic nematode species (e.g *Steinernema carpocapsae, S. feltiae* and *H. bacteriophora*) attack a wide spectrum of insects in the laboratory where host contact is assured, environmental conditions are optimal, and no ecological or behavioral barriers to infection exist. The range of insects infected by entomopathogenic nematodes after inundative field releases is considerably smaller or rather restricted. For example, *S. seapterisci*, is adapted to Orthoptera, especially mole crickets, and poorly performs in other insects. *S. kushidai* appears to be so specialized to scarabaeid larvae that it hardly infects and cannot reproduce in other hosts. The life cycle of *Steinernema* and *Heterorhabditis* is given in fig. 20.

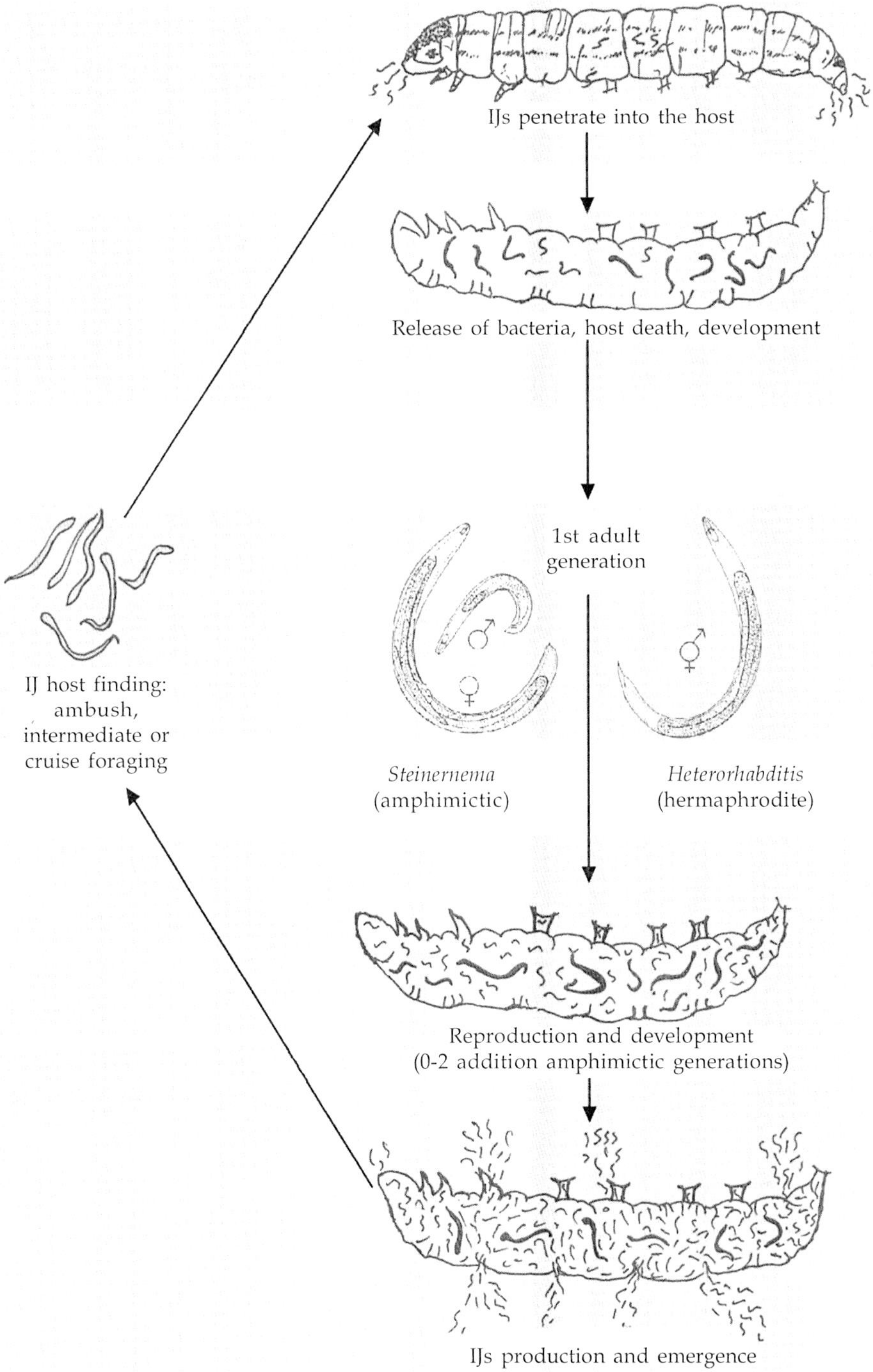

Fig. 20 : Life cycle of *Steinernema* and *Heterorhabditis*

Table 24: Described species of entomopathogenic nematode species and their respective symbiotic bacterial species

Nematode genus	Nematode species	Symbiont species
Steinernema	*affine*	*Xenorhabdus bovienii*
Steinernema	*arenarium*	*Xenorhabdus* sp.
Steinernema	*carpocapsae*	*X. nematophilus*
Steinernema	*cubanum*	*X. poinarii*
Steinernema	*feltiae*	*X. bovienii*
Steinernema	*glaseri*	*X. poinarii*
Steinernema	*intermedium*	*X. bovienii*
Steinernema	*kraussei*	*X. bovienii*
Steinernema	*kushidai*	*X. japonicus*
Steinernema	*rarum*	*Xenorhabdus* sp.
Steinernema	*riobrave*	*Xenorhabdus* sp.
Steinernema	*ritteri*	*Xenorhabdus* sp.
Steinernema	*scapterisci*	*Xenorhabdus* sp.
Steinernema	*siamkayai*	*Xenorhabdus* sp.
Heterorhabditis	*argentinensis*	*Photorhabdus luminescens*
Heterorhabditis	*bacteriophora*	*P. luminescens/ P. temperata*
Heterorhabditis	*brevicaudis*	*P. luminescens*
Heterorhabditis	*hawaiiensis*	*P. luminescens*
Heterorhabditis	*indica*	*P. luminescens*
Heterorhabditis	*marelatus*	*P. luminescens*
Heterorhabditis	*megidis*	*P. temperata*
Heterorhabditis	*zealandica*	*P. temperata*

Survival of the nematode infective juveniles applied to the soil depends on extrinsic abiotic factors like soil moisture, soil texture, temperature, relative humidity, UV radiations (*i.e.* sun light) etc. and biotic factors like natural enemies, competition and antibiosis. Direct exposure to sunlight can kill nematodes within minuets. Survival in water saturated and clay soils is less compared to well aerated sandy soil. The natural enemies like collembolans, mites, tardigrades and predatory nematodes can also predate on entomopathogenic nematodes. These entomopathogenic nematodes could be reared and mass produced on artificial chemically known solid/ liquid monoxenic media or by *in vivo* cultivation using suitable host insect. However the generally field applied/ commercial formulations that demand longer shelf life of the product at room temperature (at least six months) and ease of application over several acreage of land exist in the form of wettable dispersible granules containing infective juveniles. The species of nematodes that are being sold commercially as bioinsecticide are *S. carpocapsae, S. riobrave, S. feltiae, S. scapterisci, S. glaseri, H. bacteriophora and H. megidis.*

Advantages of the solid media process are that capital costs are low, limited expertise is required, and the logistics of production are flexible. Because of limited economics of scale, this method is mostly feasible for countries with low labour cost. Conversely, for the liquid fermentation process the proportion of labour and capital cost decreases in upscale, while operating cost increase. Liquid fermentation allows the lowest cost of mass production of entomopathogenic nematodes and is the method of choice for industrialized countries. The most common formulations of EPNS are the water dispersible granules and the most common method of application is spraying directly on to the soil (or other) surface. EPNS are safe to humans and non-target organisms and are most suitable for application to soil or cryptic (protected) habitats. Most insect targets are in the orders Coleoptera and Lepidoptera, but examples may also be found among other orders such as Blattodea, Diptera, Hymenoptera, Orthroptera, Siphonptera, and Thysanoptera. Commercially available *Steinernema* and *Heterorhabditis* species for major pests are given in table 25.

Table 25: Commercially available ectoparasitic nematode species for major pests

Commodity	Nematode Species	Common Name	Scientific Name
Citrus	Sc, Sr	Blue green weevil Sugarcane rootstalk borer	*Pachnaeus litus* *Diaprepes abbreviatus*
Fruit Trees	Sc	Stem borers	Sesiidae
Greenhouse and	Sc, Hb, Hm	Black vine weevil	*Otiorhynchus sulcatus*
Nursery Plants	Sf	Sciarid flies	Sciaridae
	Sc	Stem borers	Sesiidae
	Sc, Hb, Hm	Strawberry root weevil	*O. ovatus*
Mint	Sc	Cutworms Mint root borer Black vine weevil	Noctuidae *Fumibotys fumalis* *O. sulcatus*

Mushroom	Sf	Sciarid fly	*Lycoriella* spp.
Sugar beet	Sc	Sugar beet weevil	*Cleonus mendikus*
Turf	Sc	Armyworm Billbugs Japanese lawn cutworm	*Pseudaletia unipuncta* *Sphenophorus* spp. *Spodoptera depravata*
Vegetable and Field crops	Sc	Cutworms Cucumber beetles Flea beetles	Noctuidae Chrysomelidae Chrysomelidae

Sc- *S. carpocapsae*, Sr- *S. riobrave*, Sf- *S. feltiae*, Hb- *H. bacteriophora*, Hm- *H. megidis*

Protozoa for the Management of Insect Pests

Insects get infested and get killed by many protozoan diseases caused by chronic infection of host specific protozoan genera and thus these microbes comprise an important regulatory role in insect population. The general mode of action of these entomopathogenic protozoa is through debilitation of the overall fitness and reproduction and thereby decrease in host vigour and longevity by acting as chronic debilitating agents. Protozoa based biopesticides have not made much impact so far and not many studies and products are available. *Nosema fumiferana* is one such typical protozoan being exploited for the management of spruce budworm. *Varimorpha necatrix,* is another protozoan that infects 36 lepidopteran pests and has peculiar abilities of high virulence and wide host range making it an attractive candidate for functioning as better microbial control agent. *Nosema locustae* is a commercially produced protozoan effective against grasshoppers.

Methods of Application of Microbial Insecticides

Four different methods have been followed for application of microbial insecticides.

(i) *Introduction and colonization into an insect population*: Here the microbial insecticide from other countries are imported and introduced into a country. A nuclear polyhedrosis virus introduced from Europe to Canada successfully controlled the pine sawfly, *Neodiprion sertifer.*

(ii) *Used with parasites, predators and other pathogens*: Two pathogens put together bringing more efficient control has been reported.

The army worm, *Pseudaletia unipunctata* is rapidly destroyed due to multiple infections by nuclear polyhedrosis virus and the granulosis virus, suggesting a synergistic association. Such combination of pathogens may be useful in controlling more than one insect pathogen damaging a crop.

(iii) *Application as sprays, dusts or baits*: The commercial preparations are applied as sprays or dusts just as chemical insecticides. Promising results have been obtained with *Bacillus thuringiensis* on corn earworm, *Helicoverpa zea.*

(iv) *Used along with chemical insecticides, both compatibly and synergistically*: There are increasing numbers of reports on the compatibility of microbial insecticides with various chemical insecticides. This indicates their potential for integrated control measures. For example use of *Beauveria bassiana* and also *B. thurungiensis* with several chemical insecticides at sublethal doses resulted in effective control of insects over that of each pesticide when used alone. This also reduces residue problem.

Application methods (iii) and (iv) mentioned above are usually followed for bringing in short term control or temporary control. It is used for controlling insects of low economic threshold. Application methods (i) and (ii) mentioned above are used for bringing in long term control or permanent control. Insects having high economic threshold are controlled by pathogen on a long term basis.

Additives like flour, skimmed milk, blood albumin, vegetable oil etc. used along with the pathogen have been found to assure better application and retention on the foliage.

Time of application of microbial insecticides is more critical than that of chemical insecticides, as it has to be ingested by the insect and in some cases the pathogen requiring an incubation period within the insect body before killing the insect.

There are some advantages and disadvantages of using microbial insecticides. The advantages have been mentioned earlier. The disadvantages are

(i) The necessity for careful and correct timing of application of the pathogen with respect to the incubation period of the diseases.

(ii) The relatively marked specificity of most pathogens sometimes narrows down its use where several pests destroyed by a single chemical insecticide occur on a crop.

(iii) The necessity of maintaining the pathogen in a viable condition until the insect is contacted.

(iv) The difficulty of producing some pathogen either in large quantities or inexpensively or both, as in obligate pathogens.

(v) The tendency of dead insects to remain attached to the host plant, which is particularly objectionable with food crops.

(vi) The requirement of favourable climatic conditions for some pathogens to invade and infect the host.

Sustainable agriculture will rely increasingly on alternatives to conventional chemical insecticides for pest management that are environmentally friendly and reduce the amount of human and environmental contact with pesticides. Microbial control agents of pests, in conjunction with other integrated pest management components, can provide effective control of several insect pests.

References

Adamczyk, J.J., Adams, L.C. and Hardee, D.D. 2001. Field efficacy and seasonal expression profiles for terminal leaves of single and double *Bacillus thuringiensis* toxin cotton genotypes. *Journal of Economic Entomology*, 94:1589-1593.

Aizawa, K. 1982. Microbial control of insect pests. In: *Advances in Agricultural Microbiology* (ed. N.S.Subba Rao). Oxford & IBH Publ. Co., New Delhi, India. pp. 397-418.

Aneja, K.R. 2002. Biopesticides (Microbial pesticides) In: *Glimpses in Plant Sciences* (eds.K.R. Aneja, M.O. Charaya, A. Aggarwal, D.K. Hans and S.A. Khan). Pragati prakashan, Meerut, India. pp. 20-27.

Arora, N., Agrawal, N., Yerramilli, V. and Bhatnagar, R.K. 2007. Biology and applications of *Bacillus thuringiensis* in integrated pest management. In: *General Concepts in Integrated Pest and Disease Management* (eds. A. Ciancio and K.G. Mukerji) Springer, Dordrecht, The Netherlands. pp. 227-244.

Arora, R. and Dhaliwal, G.S. 2001. Integrated pest management: concept and approaches. In: *Microbial Control*, Kalyani Publishers, New Delhi. pp. 197-233.

Barrett, J.W., Primavera, M., Retnakaran, A., Arif, B. and Palli, S.R. 2002. Aspects of nucleopolyhydrosis pathogenesis in lepidopteran larvae. In: *Microbial Biopesticides*. (eds. O. Koul and G.S. Dhaliwal). Taylor and Francis, London, pp. 205-214.

Battu, G.S., Arora, R. and Dhaliwal, G.S. 2002. Prospects of baculoviruses in integrated pest management. In: *Microbial Biopesticides*. (eds. O. Koul and G.S. Dhaliwal). Taylor and Francis, London. pp. 215-238.

Bonning, B.C. and Hammock, B.D. 1996. Development of recombinant baculoviruses for insect pest control. *Annual Review of Entomology*. 41: 191-210.

Commandeur, P. and Komen, J. 1992. Biopesticides - options for biological pest control increases. *Biotechnology and Development Monitor*. 13:6.

Dhaliwal, G.S. and Pathak, M.D. 1993. Pesticides in the developing world: a boon in ban. In: *Pesticides: Their ecological impact in developing countries*. (eds. G.S. Dhaliwal, and B.Singh). Commonwealth Publishers, New Delhi, India.

Ebling, P.M. and Kaupp, W.J. 1997. Pathogenicity of a nuclear polyhedrosis virus to forest tent caterpillar, *Malcosoma disstria*. *The Canadian Entomologist*. 129:195-196.

Escribano, A., Williams, T., Goulson, D., Cave, R.D., Chapman, J.W. and Caballero, P. 2001. Consequences of interspecific competition on the virulence and genetic composition of a nucleopolyhedrovirus in *Spodoptera frugiperda* larvae parasitized by *Chelonus insularis*. *Biocontrol Science and Technology*. 11: 649 - 662.

Ferron, P. 1978. Biological Control of insect pests by entomogenous fungi. *Annual Review of Entomology*. 23: 409-442.

Georgis, R., Dunlop, D.B. and Grewal, P.S. 1995. Formulation of entomopathogenic nematodes. In: *Biorational Pest Control Agents, Formulation and Delivery*. (eds. F.R.Hall and J.W. Barry). American Chemical Soc. Pub., Washington, D.C., U.S.A. pp. 197-205.

Hajek, A.E., McManus, M.L. and Junior, I.D.2007. A review of introductions of pathogens and nematodes for classical biological control of insects and mites. *Biological Control*, 41:1-13.

Hussaini, S.S. and Seema Wahab. 2009. Entomopathogenic nematodes as potential biopesticides for insect pest control. Paper presented at the 5th International Conference on Biopesticides: Stakeholders Perspective, 26-30 April, 2009, India Habitat Center, Lodhi road, New Delhi, India, Sovenir. pp: 66-89.

Hussaini, S.S., Nagesh, M., Rajeswari, R. and Dar, M.H. 2007. Effect of protein and lipid sources in the standard Wout's media on the *in vitro* production and pathogenecity of *Steinernema carcocapsae* and *S. tami*. *Indian Journal of Plant Protection*. 35:93-96.

Kaaya, G.P. and Hassan, S. 2000. Entomogenous fungi as promising biopesticides for tick control. *Experimental Applied Acarology*. 24: 913-926.

Kaya, H.K. and Gaugler, R. 1993. Entomopathogenic nematodes. *Annual Review of Entomology*. 38: 181-206.

Khachatourians, G.G., Valencia, E.P. and Miranpuri, G.S. 2002. *Beauveria bassiana* and other entomopathogenic fungi in the management of insect pests. In: *Microbial Biopesticides*. (eds. O. Koul and G.S. Dhaliwal). Taylor and Francis, London. pp. 239-276.

Koppenhofer, A.M. and Kaya, H.K. 2002. Entomopathogenic nematodes and insect pest management. In: *Microbial Biopesticides*. (eds. O. Koul and G.S. Dhaliwal) Taylor and Francis, London. pp. 277-306.

Koul, O. and Dhaliwal, G.S. 2002. Microbial biopesticides. In: *Microbial Biopesticides*. (eds. O. Koul and G.S. Dhaliwal). Taylor and Francis, London. pp. 1-12.

Lacey, L.A. and Shapiro-Ilan, D.I. 2008. Microbial control of insect pests in temperate orchard systems: Potential for incorporation into IPM. *Annual Review of Entomology*. 53:121-144.

Lacey, L.A., Frutos,R., Kaya, H.K. and Vails, P. 2001. Insect pathogens as biological control agents: Do they have a future? *Biological Control*. 21: 230-248.

Menn, J.J. 1996. Biopesticides: Has their time come? *Journal of Environmental Science Health* B, 31:383-389.

Milner, R.J. 2007. Recent advances in the development of mycoinsecticides in Australia. In: *Fungi-Multifaceted Microbes*. (eds. B.N. Ganguli and S.K. Deshmukh) Anamaya Publishers, New Delhi, India. pp. 236-247.

Mishra, S. 1998. Baculoviruses as biopesticides. *Current Science*. 75: 1015-1025.

Narayanan, K. 2002. Microbial control of insect pests. Role of genetic engineering and tissue culture. In: *Microbial Biopesticides* (eds. O. Koul and G.S. Dhaliwal). Taylor & Francis Inc., New York. pp. 117-180.

Narayanan, K. 2005. Current status of Baculovirus formulations and application in India. In: *Microbial Biopesticide Formulations and Application*. (eds. R.J. Rabindra, S.S. Hussaini and B. Ramanujam). PDBC Pub., Bangalore, India. pp. 27-40.

Parente, A.F., Silva-Pereira, I. and Baldani, J.I. 2008. Construction of *Bacillus thuringiensis* wild-type S76 and Cry-derivatives expressing a green fluorescent protein: two potential marker organisms to study bacterial-plant interaction. *Canadian Journal of Microbiology*. 54: 86-790.

Parihar, A. and Siddiqui, A.U. 2004. Entomopathogenic nematodes used as biocontrol. In: *Plant Diseases - Biocontrol Management*. (ed. Sampat Nehra). Aavishkar Publishers, Jaipur, India. pp. 208-226.

Prasad, T.S.K. and Bhat, D.J. 2007. Entomogenous fungi and development of mycoinsecticides. In: *Fungi-Multifaceted Microbes*. (eds. B.N. Ganguli and S.K. Deshmukh) Anamaya Publishers, New Delhi, India. pp. 248-271.

Schwarz, M.R. 1995. *Metarhizium anisopliae* for soil pest control. In: *Biorational Pest Control Agents, Formulation and Delivery.* (eds. F.R. Hall and J.W. Barry). American Chemical Soc. Pub., Washington, D.C., U.S.A. pp. 183-196.

Sekar, V. and Sekar, P. 2006. *Bacillus thuringiensis* use in horticulture. In: *Microbial Biotechnology in Horticulture.* (eds. R.C. Ray and O.P. Ward). Science Publishers, Enfield, U.S.A. pp. 203-228.

Singh, S.P., Ballal, C. and Jalali, S.K. 1998. Production and use of nuclear polyhedrosis viruses of *Spodoptera litura* and *Helicoverpa armigera.* PDBC Pub., Bangalore.

Steinhaus, E.A. 1977. *Insect Microbiology.* Academic Press, New York, USA.

Tanada, Y. and Kaya, H.K. 1993. *Insect Pathology.* Academic Press Inc., Narcourt Brace, Javanovich Publishers, San Diego, USA.

Chapter-8

Microorganisms for Control of Weeds

Herbicides are the chemical substances used to kill weeds, *i.e.*, unwanted plants that compete with crop plants for soil nutrients. Chemical herbicides are the most effective immediate solution to most weed problems. Use of chemicals, however, can be harmful to wildlife, to the environment and to humans. Chemicals are also expensive, and weeds develop resistance to the chemicals. For these reasons, there is now a move toward alternative methods of controlling the weeds. One of the most promising is biological weed control, which involves using one living organism to control the activities of another organism.

All classes of plant pathogens *viz.* fungi, bacteria, viruses, mycoplasmas and nematodes have been considered in research and development projects to control weeds by biocontrol strategy. Among pathogens, fungi are now preferred as biocontrol agents because of the following reasons:

(i) Their ability for active invasion of plant tissues;
(ii) Easy to produce and store;
(iii) Inexpensive to use;
(iv) Easier to identify than bacteria and viruses; and
(v) Their taxonomic position is better defined.

Two approaches inoculative and inundative are used for the control of weeds.

Inundative or Bioherbicidal Strategy

Inundative method consists of application of massive doses of inoculum to a large weed population to create a fast and high level of epidemic, when conditions conducive for disease development and weed control are present using periodical releases of native control organisms to suppress weed population in an ecosystem. The formulations of native fungal pathogens used in this strategy have been referred to as mycoherbicides. Mycoherbicides are highly specific disease inducing fungi, isolated from weeds, mass produced and sprayed on fields to biologically control a specific weed without harm to the crop or any nontarget species in the environment.

Mycoherbicides have several advantages over conventional chemical pesticides. These are: (i) they are cost effective. For example in developing a new herbicide, an agrochemicals company screens up to 13,000 compounds more than half of this picked up at random, while search of mycoherbicides is more directed towards reducing the costs of development; (ii) these could also be used as alternatives to herbicides when the weed has developed resistance; (iii) these are more selective than most chemicals, making them a better choice to control weeds which are close relatives of the crops that they infect; (iv) these are not persistent in the environment unlike many chemical herbicides; (v) these are safe for humans, animals, other plants and to the environment.

Steps in the Development of a Mycoherbicide

Even though mycoherbicides are usually fungal pathogens that affect leaves and stems and already present in the infected region of the target weed, inundation of the environment with a massive dose of inoculum still requires them to be host specific, and to be effective they should still be sufficiently virulent. For the development of a mycoherbicide there are atleast seven steps which are to be followed in the sequence described. These are:

(i) Collection of basic information about the weed and the fungal pathogen.

(ii) Field survey to examine the target weed showing symptoms of infection, collecting specimens of all pathogens causing damage to the weed: The effectiveness of a plant pathogen for the control of a weed can be assessed in its original habitat by ecological observations to evaluate the relationship between changes in the weed population due to the attack of the pathogen, preferably throughout the growing season of the weed.

(iii) Studies on the biocontrol agent starting from its isolation, characterization, identification, ecology, host range etc.

(iv) Host specificity of the pathogen: A plant pathogen can only be used as a biocontrol agent of a weed when there is sufficient evidence that it will be host specific or restricted to a small number of systematically closely related weed species and which is not known to attack any economic plant in its native range.

(v) Screening for environmental safety: Fungal plant pathogens are known to produce various toxic metabolites, including mycotoxins injurious to humans and animals. Fungal spores can also be strong allergens. Keeping in view this fact, in some instances tests are to be performed to whether the candidate organism is toxic or allergic to humans and animals or capable of inducing a mammalian toxin in plants.

(vi) Determination of efficacy: Initial efficacy determinations are conducted in the laboratory and small field plots. Because of the uncertainties in translating laboratory and green house tests and results to the field, field experiments are still required to determine efficacy for commercial development of a biocontrol agent (mycoherbicide).

(vii) Formulation: Formulation is the most essential step for the successful application of pathogens for weed control. In addition to this, the availability of bioherbicide in familiar forms that are easy to apply will aid in consumer acceptance. Formulation material is generally a liquid carrier plus other component (such as adjuvants or surfactants) which when combined with pathogen spores or propagules, produce a suspension solution that can be efficiently applied to target weed.

Few pathogens commercially produced and presently used as mycoherbicides are *Phytophthora palmivora* against stranglervine, *Morrenia odorata*; *Colletotrichum gloeosporioides* f.sp. *aeschynomene* against northern jointvetch, *Aeschynomene virginica*; *Alternaria cassiae* against *Cassia obtusifolia*; *Colletotrichum gloeosporioides* f.sp. *malvae*, against *Malva pusilla*; *Chondrostereum purpureum* against *Prunus serotina*. Few others showing promising results are *Colletotrichum coccodes* against the velvet leaf, *Abutilon theophrastii*; *Colletotrichum gloeosporioides* f.sp. *cuscutae* against dodder, *Cuscuta* sp.; *Cercospora rodmanii* against water hyacinth, *Eichhornia crassipes*; *Phomopsis convolvulus* against field bindweed, *Convolvulus arvensis* and *Cercospora partheniphila* against the congress weed, *Parthenium hysterophorus*. For the control of weedy tree species, for example, the red alder *Alnus rubra* in a forest, the bioherbicide *Chondrostereum purpureum* has been applied to wounded stems or stumps to inhibit resprouting by enhancing decay of the woody tissue. Soil-borne fungi can be important bioherbicide candidates since these fungi applied directly to soils can reduce weed populations through decay of seeds prior to emergence or kill seedlings shortly after emergence.

Plant pathogenic bacteria including *Xanthomonas campestris* pv *poannua* (Xcp) and *Pseudomonas syringae* pv *tagetis* (Pst), have also been developed as bioherbicides for control of annual bluegrass (*Poa annua*) and Asteraceae (composite) weeds, respectively. The Xcp bioherbicide must be applied by spraying the bacterial suspension while mowing to allow bacterial cells to invade wounded tissue of the grass. The Pst herbicide is prepared with an organosilicone surfactant to enhance bacterial infection of leaf and stem tissue on onset of disease.

Another group of bacteria under intensive investigation for bioherbicidal potential are deleterious rhizobacteria (DRB), which differ from bacterial pathogens in that they are non-parasitic bacteria colonizing plant roots and able to suppress plant growth without invading the root tissues. *Pseudomonas fluorescens* D7 is a soil applied DRB bioherbicide formulated as a liquid suspension or encapsulated in clay that effectively suppresses downy brome (*Bromus tectorum*) in cereal grain crops. Microorganisms with the potential to be used for the biocontrol of weeds are given in table 26.

Table 26: Microorganisms effective for the biocontrol of weeds

Agent	Target weed	Ecosystem
Foliar/stem fungal pathogens		
Puccinia spegazzini	Mikania weed (*Mikania micrantha*)	Forests, plantation crops
Colletotrichum gloeosporioides f.sp. *aeschynomene*	Northern jointvetch (*Aeschynomene virginica*)	Rice, soybean
Colletotrichum gloeosporioides f.sp. *malvae*	Round-leave mallow (*Malva pusilla*)	Flax, lentils, strawberries
Phytophthora palmivora	Strangler or milkweed vine (*Morrenia odorata*)	Citrus groves
Colletotrichum truncatum	Hemp sesbania (*Sesbania exaltata*)	Soybean, cotton rice
Colletotrichum coccodes	Velvetleaf (*Abutilon theophrasti*)	Maize, soybean
Puccinia canaliculata	Yellow nutsedge (*Cyperus esculentus*)	Horticulture crops
Colletotrichum gloeosporioides f.sp. *cuscutae*	Dodders (*Cuscuta* spp.)	Ornamental plants and trees
Chondrostereum purpureum	Black cherry (*Prunus serotina*) Red alder (*Alnus rubra*)	Forest
Nectria ditissima	Red alder	Forest
Cercospora rodmanii	Water hyacinth (*Eichhornia crassipes*)	Aquatic sites
Soil-borne fungal pathogens		
Rhizoctonia solani	Leafy spurge (*Euphorbia esula*)	Rangelands
Non-phytopathogenic soil-borne fungi		
Gliocladium virens	Several	Horticultural crops
Foliar bacterial pathogens		
Pseudomonas syringae	Canada thistle (*Cirsium arvense*) Cocklebur (*Xanthium* sp.)	Maize, soybean
Non-pathogenic bacteria		
Pseudomonas fluorescens D7	Downy brome (*Bromus tectorum*)	Cereal grains

One of the commercially produced and extensively used mycoherbicide COLLEGO is discussed here. Curly indigo or northern jointvetch (*Aeschynomene virginica*), a black seeded leguminous plant is a serious weed in rice fields in the USA. The seeds of this weed are difficult to remove from harvested grain. In 1969 an anthracnose disease of curly indigo, infecting leaflets, petioles, stems, seedpods and seeds, at endemic levels was discovered. The causal organism was isolated and identified as *Colletotrichum gloeosporioides* f.sp. *aeschynomene*. An understanding of anthracnose disease etiology gained from the literature and continued by empirical studies with the weed and fungus revealed that the pathogen was specific and potentially devastating to its weed host, but was constrained in natural situations by poor capacity for dissemination of inoculum. Host specificity tests were conducted at green house and field levels. Protocols for testing were developed with the co-operation and approval of the U.S. Environmental Protection Agency (EPA) which issued Experimental Use Permits (EUP) for large scale field tests from 1975 through 1981. Safety of the fungus to animals was confirmed in 1975 by the toxicology tests. Animals including rats, mice, rabbits, guinea pigs, dogs, turkeys, chicken, quail, ducks, crayfish, perch, catfish, frogs and earthworms were tested *via* oral, nasal, dermal, occular, or interperitoneal injection. Autopsies were conducted in challenged animals and on animals challenged with normal saline solution (control). No infection and other serious tissue effects were observed in animals treated with the fungus, and the fungus was killed by the digestive system of the rat. Feeding of the culture broth after growth of the fungus likewise revealed no harmful effects, indicating the absence of mycotoxin production by the fungus. Large scale field tests, which were initiated in 1973 continued for 8 years. Optimum control of curly indigo resulted when fresh spores of the fungus were applied at the rate of 187 billion spores per hectare in a volume of 94 litres/hectare. Weed plants exhibited symptoms after one week and were killed within five weeks of application. Weed kill typically ranged between 95 and 100%.

Technology for drying and formulating spores on a commercial scale was developed. Dry, living spore formulations were field tested with cooperative growers. Experiments on longevity and death of the fungus after application together with extensive laboratory tests with plants and animals, assured EPA of the safety of the formulation and led eventually to formal approval for COLLEGO. COLLEGO is a wettable powder formulation consisting of 15% viable spores and 55% inert ingredients. A single application of COLLEGO in the growing season of the crop, when the curly indigo plants emerge through the rice, kills the weed in five weeks.

Another example of a classical biological control (CBC) of weeds worth monitoring is the host-specific plant pathogen, *Puccinia spegazzinii* against mikania weed (*Mikania micrantha*) in India. The neotropical vine mikania weed is an increasing threat to natural and man-made forests as well as to several agricultural and horticultural ecosystems in India. Although in its native range *M. micrantha* is rarely weedy, in its exotic range, especially in south and south-east Asian countries, it has become an intractable weed over the past several decades. Because of its rapid growth habit, the plant, which can smother even such hardy trees as teak, eucalyptus, rubber, oil palm and cocoa, has acquired one of its common names, mile-a-minute weed. Whereas in north-east India it is a great problem particularly in tea, it is an equally big problem in plantation crops in south-west India. The common control measures that are prevalent in tea gardens and plantations are either cultural or chemical means. These methods are expensive and impracticable. Moreover, chemical herbicides can be very harmful to the non-target plants, people and the environment.

Between 1996 and 1999, surveys were conducted throughout the tropical and sub-tropical American native range of *M. micrantha* for pathogens having potential for CBC in India. Three microcylic rust species, *P. spegazzinii, Dietelia portoricensis* and *Dietelia mesoamericana,* were found to occur in association with the plant. However, *P. spegazzinii* was selected for use as a CBC agent in India after an extensive host-range screening and studies on the environmental requirements for the fungus. This rust is microcyclic, autoecious species that infects all aerial parts of the plant causing necrosis and cankering, leading to plant death (Plate 23).

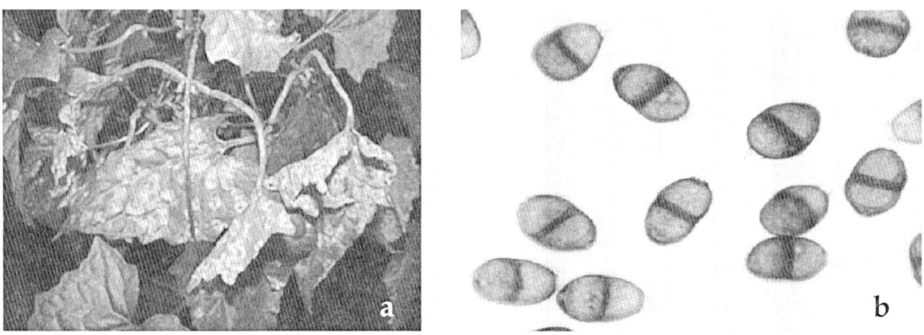

Plate 23 : a. *Mikania micrantha* infected with *Puccinia spegazzinii* b. Spores of *P. spegazzinii*

These studies were carried out in the CABI- Europe-UK quarantine in Ascot with funding from DFID, under a collaborative project between CABI and research institutions in India between 1996 and 2000.

P. spegazzinii was imported into National Containment-cum-Quarantine Facility (CQF) for Transgenic Planting Material of the National Bureau of Plant Genetic Resources (NBPGR) in New Delhi during 2003 and 2004. After establishing the fungus at NBPGR, an additional host-specificity screening was undertaken during 2004 and 2005. This involved 74 plant species, including 18 species that were earlier tested in the UK, and reconfirmed the results from the UK: the rust is totally specific to *M. micrantha.* At NBPGR, the rust was found to be pathogenic to population of mikania weed from several locations within Kerala and Assam, which indicated that *P. spegazzinii* has considerable potential as a CBC agent for mikania weed in India.

A supplementary Dossier on the additional host-specificity tests provided the basis for obtaining the permit for release of *P. spegazzinii* from the Plant Protection Advisor of the Government of India, Ministry of Agriculture in June 2005. 'Limited' field releases of the rust have been made since 2005 in both Assam and Kerala. This is the first time that a fungal pathogen is being used as a CBC for mikania weed.

The mikania weed CBC project in India has become a 'flap-ship' project. Other Asian countries, including China, are following the Indian example for future management of *M. micrantha* using *P. spegazzinii.*

Some of the important weeds in India on which research is needed are *Ageratina adenophora* (croton weed), *Chromolaena odorata* (chromolaena weed), *Cyperus rotundas* (nut grass), *Eichhornia crassipes* (water hyacinth), the invasive grasses *Echinochloa crus-galli* (barnyard grass), *E. colona* (jungle rice) and *Phalaris minor* (littleseed canary grass), *Lantana camara* (lantana weed), *Mimosa diplotricha* (the giant sensitive plant) and phanerogamic parasitic weeds like *Striga* spp. on cereals and legumes, *Orabanche* spp. on tobacco and sunflower and *Cucuta* spp. on ornamentals and trees.

Mycoherbicides are influenced greatly by environmental conditions following application to target weeds. For foliar pathogens, temperature, free moisture and protection from ultra violet irradiation often are critical to plant infection. In particular, most foliar fungal pathogens require several hours of free moisture on plant surfaces for spore germination, infection structure formation and plant infection. Once the pathogen successfully penetrates the outer plant tissues and establishes a parasitic relationship with the host, water is supplied by the host cells and is no longer critical. In many geographic locations, free moisture in the form of dew or rainfall often is not sufficient and is a major constraint to consistent performance of biological agents.

Formulation of a biological to delay evaporation of the application spray or reduce the free moisture requirement of the agent often is a key component of successful biological control. Water-in-oil, or invert, emulsions have been used successfully to trap water around fungal spores and retard water loss by evaporation. Several invert formulations composed of paraffin, paraffin oil and an emulsifying agent have been used to eliminate or significantly reduce the free moisture requirement of several fungal weed control agents. In addition to minimizing environmental constraints, formulation can be used to increase the efficacy of a biocontrol agent by enhancing pathogen infectivity or by increasing the susceptibility of the target host. Formulation components can enhance pathogen infectivity by increasing spore viability, spore germination, infection structure development or metabolic activity associated with plant infection and disease development. Vegetable oil adjuvants, fatty acid monomers used during formulation were found to enhance virulence of the biocontrol agent.

More research is needed in use of microorganisms for control of weeds in cropping systems. Bioherbicides will be of significant value in managing weeds in areas where herbicides are not effective due to regulations that severely restrict or prohibit herbicide use and where preservation of the environment is the primary goal. These special situations include restoration of native ecosystems, wetlands, national parks, wildlife refuges and areas bordering waterways.

References

Aneja, K.R. 1997. Discovery and development of mycoherbicides for biological control of weeds. In: *New Approches in Microbial Ecology*. (eds. J.P. Tewari, G. Saxena, N. Mittal, I. Tewari and B.P. Chamola). Aditya Books Private Ltd., New Delhi, India, pp. 517-555.

Aneja, K.R. and Khan, S.A. 2002. Congress grass (*Parthenium hysterophorus* L): Its impact, infestation and biocontrol with fungal pathogens - An overview. In: *Glimpses in Plant Sciences* (eds. K.R. Aneja, M.U. Charaya, A. Aggarwal, D.K. Hans and S.A. Khan). Pragati Prakasthan, Meerut, India, pp. 317-332.

Aneja, K.R., Charaya M.U., Aggarwal, A., Hans, D.K. and Khan, S.A. Biopesticide (Microbial Pesticides). In: *Glimpses in Plant Sciences*. (eds. K.R. Aneja, M.U. Charaya, A. Aggarwal, D.K. Hans and S.A. Khan). Pragati Prakashan, Meerut, India. pp. 20-27.

Bishop, A. 2000. Bioherbicides: New products and new models of action. *Tas-Regions*. 6: 41-42.

Dyke, C.G.V. 1991. Biological control of weeds with fungi. In: *Handbook of Applied Mycology*. Vol.1. *Soil and Plants*. (eds. D.K. Arora, B. Rai, K.G. Mukerji, and G.R. Knudsen). Marcel Dekker, New York, USA, pp. 357-378.

Green, S. 2003. A review of the potential for the use of bioherbicides to control forest weeds in the UK. *Forestry*. 76: 285-298.

Hallet, S.G. 2005. What are the bioherbicides? *Weed Science*. 53: 404-415.

Kremer, R.J.2002. Bioherbicides: Potential successful strategies for weed control. In: *Microbial Biopesticides*. (eds. O. Koul and G.S. Dhaliwal). Taylor and Francis, London, UK. pp. 307-323.

Pandey, A.K., Gayathiri, S., Rajak, R.C. and Hasija, S.K. 1996. Possibilities, problems and prospects of microbial management of *Parthenium hysterophorus* in India. In: *Perspectives in Biological Sciences*. (eds. V. Rai, M.L. Naik, C. Manoharachary). Dazzle offset printers, Raipur, India.

Rees, N.E., Quimby, P.C., Coulson, J.R. 1995. Biological weed control strategy: An overview. In: *Biorational Pest Control Agents, Formulation and Delivery*. (eds. F.R. Hall and J.W. Barry) American Chemical Society Pub., Washington, D.C., USA. pp. 252-269.

Sreerama kumar, P., Rabindra, R.J. and Ellison, C.A. 2008. Expanding classical biological control of weeds with pathogens in India: the way forward, XII International Symposium on Biological Control of Weeds. pp. 165-172.

Tewari, J.P., Geeta Saxena., Neelima Mittal., Ishwari Tewari and Chamola, B.P. 1997. Discovery and development of mycoherbicides for biological control of weeds. In: *New Approches in Microbial Ecology*. Aditya Book Pvt. Ltd., New Delhi, India. pp. 517- 555.

Weidemann, G.J., Boyette, O.D. and Templeton, G.E. 1995. Utilization criteria for mycoherbicides In: *Biorational Pest Control Agents, Formulation and Delivery*. (eds. F.R. Hall and J.W. Barry) American Chemical Society Pub., Washington, D.C., USA. pp. 238-251.

Chapter-9

Microorganisms in Conversion of Agricultural Waste to Compost

Sustainable agriculture or organic farming wherein biological inputs are maximum and chemical inputs are minimal, is drawing the attention of farmers across the world including India. For replacing chemical fertilizers, which supply plant nutrients by organic manure, the basic requirement is the availability of raw material *i.e.* the organic matter or biomass that can be decomposed to organic manure. The question that is often asked is what are the different sources of available biomass and how much of it is available? Different sources of biomass can be categorized as

(i) Livestock and human wastes: cattle shed wastes, other livestock and human excreta, slaughter house waste, piggery and poultry wastes.

(ii) Crop residues, tree wastes and aquatic weeds: cereal residues, sugarcane trash, forest/tree litter, aquatic weeds, groundnut shells etc.

(iii) Green manures: *Gliricidia maculata, Pongamia glabra, Calotropis gigantea, Tephrosia purpurea, T. candida, Indigofera teysmanii, Cassia tora, Sesbania speciosa, Ipomoea carnea* etc.

(iv) *Urban and rural wastes*: solid wastes, sewage and sullage, biogas slurry, kitchen wastes etc.

(v) *Agro-industrial wastes*: press mud, bagasse, paper and pulp wastes, oil cakes, paddy bran and husk, coir pith, sawdust etc.

(vi) *Marine wastes*: seaweeds, fishmeal etc.

A careful cataloguing of wastes accumulating in India every year reveals plentiful of substrate. Crop residues from rice, maize, sugarcane, sorghum, pearl millet, pulses and biomass from aquatic weeds and from wasteland weeds if properly processed can supply 2.20 million tonnes of fertilizers equivalent as NPK. With all these resources there is still a constraint in the availability of organic matter for composting. Cow dung is burnt as fuel and most of the crop residues are used as fodder for animals or for making chip-board etc. Under these circumstances agro-based industrial wastes like press mud and spentwash etc. can be profitably used for producing organic manure.

Composting is one of the oldest methods of preparation of organic manure and has developed into a key technology for disposal and recycling of rural and urban organic wastes. Composting has been the basis for organic methods of farming and gardening. It was shown that best compost was prepared when crop residues and cattle dung are mixed together in the ratio of 3:1. Various organic compounds which reach soil by way of animal and plant residues are made up of complex carbohydrates, simple sugars, starch, cellulose, hemicelluloses, pectins, gums, mucilage, proteins, fats, oils, waxes, resins, alcohols, aldehydes, ketones, organic acids, lignin, phenols, tannins, hydrocarbons, alkaloids, pigments and other products. The organic sources used for composting, in general, contain 15-60% cellulose, 10-30% hemicellulose, 5-30% lignin, 5-40% proteins, 5-13% minerals (ash) beside 2-30% of cold water soluble fraction (sugars, starch, amino acids, urea, ammonium salts) and 1-15% of ether and alcohol soluble fats, oils and waxes. Cellulose consists of unbranched polymers of β-linked glucose residues arranged in linear chains, where every other glucose residue is rotated approximately 180°. As a result cellobiose is the structural repeating unit of the glucan chains in cellulose. The structure of cellulose appears to be controlled by the biosynthetic machinery. At least six different crystalline forms of cellulose exist, but higher plants synthesize normally only two. They are closely related forms of cellulose - I, namely Iα and Iβ.

Hemicelluloses has been widely used to refer to those cell wall polysaccharides normally found in association with cellulose but soluble in dilute alkali. The hemicelluloses are second only in quantity to cellulose and represent a significant source of energy and nutrients to the microflora. Most of the hemicelluloses are found in close physical proximity to cellulose in primary and secondary cell walls of higher plants. Straw and bark consist up to 30% hemicellulose, sugarcane bagasse has 30%, conifer wood has 7-12% hemicellulose and deciduous wood has 20-25% hemicellulose. They function as storage and supporting substance in plants. Xylans are the predominant constituents of hemicelluloses but mannans, glucomannoglycans, galactans, arabinogalactoglycans, arabinoxyloglycans and other polysaccharides are also included in hemicelluloses. Pectin is an example of hemicelluloses and is an important constituent of the middle lamella of the cell walls. Pectin is degraded by enzyme pectinase, which is a complex of several enzymes. Lignin is much more complex than celluloses and is formed by chemical reaction involving phenols and free radicals without any specific order. Lignin gets encrusted on the cellulose and hemicellulose matrix. The lignin molecule has only three elements, carbon, hydrogen and oxygen. The molecule is a polymer of aromatic nuclei with either a single repeating unit or several similar units as building blocks. The repeating units range from about 200-1,000 depending upon the origin of lignin and the methods used to determine the molecular weight. Microorganisms act on each of these components and degrade them to a form that can be taken up by plants as nutrients (Table 27). Hence, the Nobel Laureate Selman A. Waksman said, "Without microorganisms, life on this planet would stand still".

Table 27: Predominant microorganisms that degrade the organic constituents

Substrate	Microorganisms	Genera of Microorganisms
Cellulose	Bacteria	*Bacillus, Cellulomonas, Cytophaga, Clostridium, Corynebacterium, Pseudomonas, Vibrio, Achromobacter, Angiococcus, Cellfalcicula, Polyangium, Sporocytophaga*
	Fungi	*Aspergillus, Alternaria, Chaetomium, Coprinus, Fomes, Fusarium, Myrothecium, Penicillium, Polyporus, Rhizoctonia, Rhizopus, Trichoderma, Trametes,*

		Trichothecium, Verticillium, Agaricus, Zygorynchus, Acremonium, Irpex, Phanerochaete, Schizophyllum, Sporotrichum, Talaromyces, Thielavia
	Actinomycetes	*Streptomyces, Nocardia, Micromonospora, Streptosporangium*
Lignin	Bacteria	*Pseudomonas, Flavobacterium, Arthrobacter, Cornyebacterium, Micrococcus*
	Fungi	*Agaricus, Armillaria, Clavaria, Clitocybe, Coprinus, Fomes, Ganoderma, Marasmium, Panus, Pluerotus, Polyporus, Pori, Collybia, Flammula, Hypholoma, Lepiota, Mycena, Pholiota, Arthrobotrys, Cephalosporium, Humicola, Phanerochaete, Volvariella, Pycnoporus, Aspergillus, Polystictus, Stereum, Trametes, Ustulina*
	Actinomycetes	*Nocardia, Thermomonospora*
Hemicellulose	Bacteria	*Bacillus, Cytophaga, Erwinia, Pseudomonas, Acromobacter, Sporocytophaga, Lactobacillus, Vibrio,*
	Fungi	*Aspergillus, Alternaria, Penicillium, Fusarium, Trichoderma, Chaetomium, Trichothecium, Rhizopus, Zygorynchus, Helminthosporium, Coriolus, Fomes, Polyporus, Glomerella*
	Actinomycetes	*Streptomyces*
Starch	Bacteria	*Bacillus, Clostridium, Cytophaga, Flavobacterium, Micrococcus, Pseudomonas, Achromobacter, Chromobacterium*
	Fungi	*Aspergillus, Fomes, Fusarium, Polyporus, Rhizopus*
	Actinomycetes	*Micromonospora, Streptomyces, Nocardia*
Pectin	Bacteria	*Bacillus, Clostridium, Pseudomonas*
	Fungi	*Fusarium, Verticillium*
Chitin	Bacteria	*Pseudomonas, Chromobacterium, Cytophaga, Flavobacterium, Clostridium, Micrococcus, Achromobacter, Beneckea*
	Fungi	*Mortierella, Verticillium, Paecilomyces, Gleomastix, Fusarium, Mucor, Trichoderma,*

		Aspergillus, Gliocladium, Penicillium, Thamnidium, Absidia
	Actinomycetes	*Actinoplanes, Micromonospora, Nocardia, Streptomyces, Streptosporangium*
Inulin	Bacteria	*Beneckea, Micrococcus, Clostridium, Cytophaga, Pseudomonas, Flavobacterium*
	Fungi	*Aspergillus, Penicillium, Fusarium*
Proteins and nucleic acids	Bacteria	*Pseudomonas, Bacillus, Clostridium, Serratia, Micrococcus, Bacillus*
	Fungi	*Rhodotorula, Mortierella, Penicillium*
	Actinomycetes	*Streptomyces*
Cutin	Bacteria	*Bacillus*
	Fungi	*Rhodotorula, Mortierella, Penicillium*
	Actinomycetes	*Streptomyces*
Tannin	Fungi	*Penicillium, Aspergillus*
Humic Acid	Fungi	*Penicillium, Polystictus*
Fulvic acid	Fungi	*Poria*

Composting

Composting can be defined as a process of bioconversion of complex organic substances by the activities of a mixed microbial population into simpler, readily available nutrient substances under optimum conditions of moisture and aeration. Composting is a microbiological, non-polluting and safe method for disposal and recycling of organic wastes by bioconversion to manures. It is a solid waste management method wherein the organic component of the wastes is biologically decomposed and stabilized under controlled conditions to a state where it can be handled, stored and applied to supply essential plant nutrients without affecting the environment. Staggering volumes of organic wastes are generated everyday by the human / industrial activity. They are nuisance at the point of generation. Also they cannot be used due to their phyto-toxicity, nitrogen immobilization and perhaps structural incompatibility. Composting reduces the odours, reduces difficulties in handling and inactivates pathogens and weed seeds, thus, providing uniform organic material suitable for soil application. The benefits of compost/ composting, thus, range from recycling of locally available agro industrial wastes for hygienic disposal, reduction in the bulkiness of the material, conservation of plant nutrients, elimination of human pathogens and mitigation of environmental pollution.

Bioconversion of organic materials is carried out by different groups of heterotropic microorganisms such as bacteria, fungi, actinomycetes and protozoa. During the course of composting, both qualitative and quantitative changes occur in the active flora, some species multiply rapidly at the first stage, change the environment and then disappear to allow other population to succeed them. Microorganisms utilize organic matter (which comprises living or dead bodies of plants, animals and microorganisms) mainly to derive their carbon and energy sources in addition to obtaining other nutrients to synthesize their own cellular components for their growth and reproduction.

Composting process can be basically divided into four phases, namely:

(i) Mesophilic or initial low active phase during which the readily available constituents like simple sugars and some organic components are used up by mesophilic microorganisms leading to rise in temperature; fungi and acid-producing bacteria appear during this stage. This phase is characterized with a temperature range of 20° to 40° C. Yeasts show profound activity directly proportional to the amount of sugars present in the composting mix. It is extremely difficult to assign the percentage composting done by bacteria or fungi. However, bacterial population plays a major role, at least in the initial phase and the fungal population predominates the later phase.

(ii) Thermophilic phase in which the temperature is high encouraging thermophilic microorganisms which degrade more complex/ recalcitrant constituents of the organic matter. This phase starts with an increase in temperature from 40°C and goes up to 70°C, which establishes predominance of thermophiles, eradicates pathogens and expedites composting.

(iii) Cooling phase is wherein the mesophilic phase with mesophilic microorganisms are restored; this is characterized by the decrease in temperature below 40°C. During subsequent process of cooling and maturing as in table 28 there is abundance of nitrogen fixers, sulphur oxidizers, hydrogen oxidizers and nitrifiers, thus, products of extracellular polysaccharides are comparatively more in this phase.

Table 28 : Predominant organisms in the different phases of composting

Microorganism	Mesophilic phase	Thermophilic phase	Cooling and Maturation phase
Bacteria	*Cellulomonas* sp. *Corynebacterium* sp. *Arthrobacter* sp. *Pseudomonas* sp. *Bacillus* sp. *Myxococcus* sp.	*Bacillus subtilis* *B. stearothermophilus* *Thermus thermophilus* *Clostridium thermocellum*	*Thiobacillus* sp. *Azotobacter* sp. *Azospirillum* sp. *Beijerinckia* sp. *Derxia* sp. *Acetobacter* sp. *Klebsiella* sp. *Beggiatoa* sp.
Fungi	*Penicillium* sp. *Aspergillus* sp. *Trichoderma viride* *Trichurus spiralis* *Paecilomyces* sp.	*Absidia* sp. *Humicola januginose* *Thermoascus aurantiacus* *Humicola insolens* *Chaetomium thermophile* *Geotrichum candidum* *Aspergillus fumigatus* *Mucor pusillus* *Torula thermophila* *Humicola griseus*	*Penicillium* sp. *Aspergillus* sp.
Actinomycetes	*Streptomyces* sp. *Nocardia* sp. *Micromonospora* sp.	*Thermoactinomyces* sp. *Thermomonospora* sp. *Micromonospora vulgaris* *Nocardia brasiliensis* *Pseudonocardia thermophila* *Thermopolyspora* sp. *Streptomyces thermofuscus* *Streptomyces thermoviaceus* *Streptomyces thermophilus*	*Streptomyces* sp. *Nocardia* sp.

(iv) Maturity/stabilization phase is with least microbial activity as most of the microbially degradable substances have already been utilized by microorganisms. This phase is initiated with decrease in temperature; the quality of compost increases due to continuous alteration in the nature of humus, as there is increase in humic acid content and decrease in fulvic acid content. Deep reddish brown and easy fragility of the compost and enhanced water holding capacity are the first visual criteria of maturation/ stabilization of compost. The temperature fluxes during composting are depicted in fig. 21.

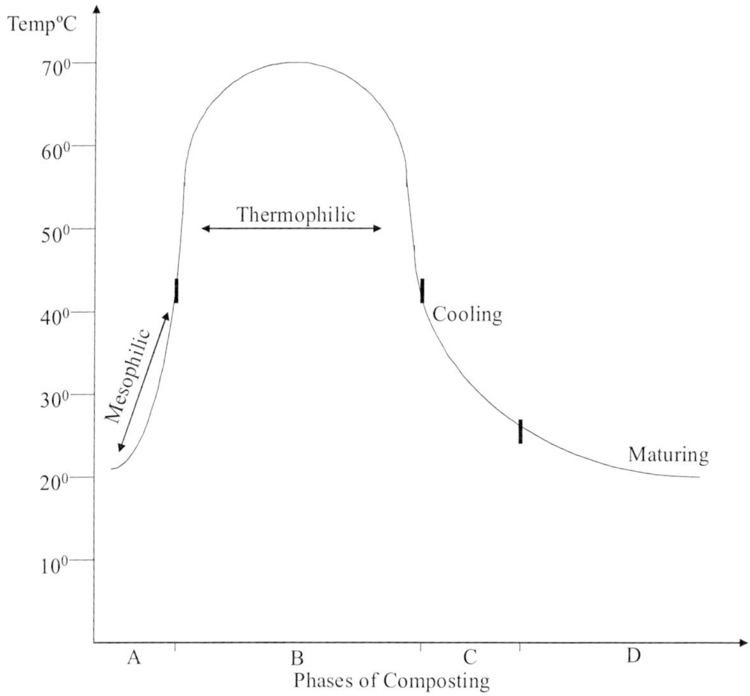

Fig. 21 : Temperature fluxes during different stages of composting

Mechanism of Action

Approximately, 50% lignin and some hemicellulose are partially degraded during composting. Three-dimensional structure of lignin (phenylpropanoids) is considerably resistant to microbial degradation; the degradation of lignin is done by fungi, which hastens solubilization and detoxification. The byproducts of hemicellulose degradation provide nutrition to the microorganisms. Lignin degradation is a prerequisite for cellulose degradation, which in turn removes barriers for composting. Amorphous forms of cellulose are degraded at a faster rate as compared to crystalline form of cellulose.

The final product of decomposition of organic matter is CO_2, water and humus. Humus is the dark amorphous material, which consist of undecomposed organic matter, which is resistant for degradation along with dead microbial cells plus plant nutrients in water soluble form. This is the material, which is incorporated to soil as organic manure to provide plant nutrients and also to improve physical conditions of the soil. Humus can be defined as a ligno-protein complex or an amino acid-lignin complex containing approximately 45% lignin compounds, 35% amino acids, 11% carbohydrates, 4% cellulose, 7% hemicellulose, 3% fats, waxes and resins and 6% other miscellaneous substances including plant growth promoting substances and inhibitors.

Further, biodegradation products of lignin are closely linked with those of humus substances. Fulvic acid, humin and humic acid have been recognized as the three major fractions by subjecting humus complexes to solvent extraction procedures. Fulvic acid is alkaline as well as acid soluble part of soil organic matter and contains carbohydrates and proteins. Humin is resistant to cold alkali and is chemically heterogeneous fraction. Humic acids form the bulk of humus complex and are regarded as polymers of aromatic compounds. A wide variety of phenolic degradation products can be obtained from humic acids.

Methods of Composting

There are different methods of composting. Basically the process is carried out either in pits or heaps. In the heap system, decomposition is carried out 1-2 m above ground. This consists of different configurations. The pile refers to a simple process caused by dumping of wastes into a cone or pyramidal shape on the ground. These may or may not be mechanically turned. The heap is same as pile but is a more indiscriminately dumped and of larger mass. Windrows are piles, turned using various machines. Large scale composting at out doors involves windrow method. They vary in size depending on the type of turning equipment utilized. The height of the windrows ranges from 1.5 to 3.0 m, with a width ranging from 2 to 3 m. Composting process is continued in windrows for 3 months and later placed in curing pile for four to six weeks depending on the substrate. The product formed can be utilized by selling or distributed in the society.

Different methods of composting have been proposed and used, altering the methodology for convenience and for faster degradation.

Bangalore method

This method of composting was developed by Acharya in 1939. This method is suitable in places where night soil and refuse are used for preparing the compost and in areas having scanty rain fall. Here composting is carried out in trenches; their size vary based on population; a population of more than 50 can have trench dimension of 10 m x 2.5 m x 1 m, length, breadth and depth respectively. The trenches should preferably have sloping walls and a floor of 90 cm slope to prevent waterlogging. A layer of refuse about 15 cm depth is spread at the bottom of trench, over this night soil is added at 5 cm, similarly like this alternative layers are made till about 30 cm above the ground level. Finally the pit is covered with a 15 to 20 cm thick layer of refuse. The materials are allowed to remain in the pit without turning and watering for 90 days. During this duration, there is reduction in the volume of biomass and we can add additional layer of night soil and refuse in alternative layers on the top, and this is plastered with mud to prevent loss of moisture and breeding of flies. Initially, there is aerobic decomposition and the temperature rises to 60-70°C, then this is taken over by anaerobic digestion occurring at the slower rate for a period of 180-240 days to obtain end product, which is odourless and has high manurial value.

Indore method

This method is suitable when there is mellifluous plant residues, animal dung, urine, wood ash etc. The plant wastes include vegetable wastes, weeds, stalks, stems, fallen prunings, chaff, fodder remnants, green matter etc. Here the total composting period taken is about three months.

Pit method

The site selected for pit method should be near cattleshed and at a high level so that no rainwater gets in during the monsoon season. Pits are dug at about 1 m deep and 1.5 - 2.0 m wide and of any suitable length. The material got from cattleshed like bedding material and other wastes are mixed and spread in the pit, each layer of the substrate is covered with slurry of dung made with 4.5 kg urine-earth and 4.5 kg of inoculum taken from a 15-day-old compost pit. The pit is sufficiently moistened and care should be taken to avoid compaction of the substrate. The substrate in the pit is turned after 15 days, 30 days and 60 days from the date of filling the pit. During each turning, the substrate is properly mixed and sufficiently moistened.

Heap method

Heap method is preferred in places where there is low rainfall. The site for heap preparation is selected at an elevation. The pile should have a dimension of 2 m long, 1.5 m high, 2 m wide and the top of 0.5

m narrowed by tapering. A layer of carbonaceous materials such as leaves, hay straw, saw dust, wood chips, corn stalks etc. are spread to about 20 cm height, this is then covered by nitrogenous materials like grass, weeds, garbage, other plant residues, manure, digested sewage sludge etc. This layering is repeated similarly till the pile is 1.5 m high and moistened. The heaps are turned at 6 and 12-week interval.

NADEP method

This method is in parallelism with heap method of composting. The NADEP method employs a brick lined tank of size about 1 x 2 x 4 m. Aeration is achieved by holes on all sides of the brick tank, helping air circulation. Common substrates employed in this method are agro wastes *viz.*, cattle feed refuse, weeds, grasses, leaves, sugarcane trash, stalks, roots, plant clippings etc. Methodically the substrates consisting of organic wastes, cow dung slurry and soil are layered; at the bottom, a layer of 150 kg organic wastes are spread, on this cow dung slurry (6 kg cow dung in 100 litre water) is spread, a third layer consisting of 60 to 80 kg cleaned sieved dry soil is spread and over it water is sprinkled. The layering is repeated similarly up to about 1 m. Finally a layer of cow dung soil paste is added. Aerobic decomposition process takes about 90 to 120 days.

Japanese method

In this method instead of pits, vats of stone slabs are made which measures 5.5 - 9.0 m in length and 0.9 - 1.2 m in width and have a height of 0.75 - 1.5 m. The stone walls are laid in such a way that aeration can be made. The stones can also have holes. At the bottom of the vat, the stone slabs are laid and plastered with cement to prevent leaching of the nutrients. This method employs five different substrates *viz.*,

(i) Cow dung and poultry wastes.

(ii) Grass, crop residues and weeds that include fresh and dry green leaves.

(iii) Wastes from agro-based industries like press mud, sericulture wastes, wastes from vegetable and fruit processing.

(iv) Naturally occurring nutrient source like rock phosphate.

(v) Organic wastes from towns and cities like sewage and sludge.

In the bottom of the vats, a layer of highly lignaceous substrates like coconut leaves, coconut shells and other fibrous materials are layered up to 10 - 15 cm height. The second layer is spread with dried leaves, grass, groundnut shells etc. up to 10-15 cm height. These layers absorb the moisture and nutrients leached from the top. The layers after being laid are sprinkled with cow dung slurry or with bio-gas spent slurry. The third layer is made up of green leaves of pongamia, green grass, weeds and other crop residues rich in nitrogen. The fourth layer contains organic waste rich in phosphorus like roots of legume crop, green manures and rock phosphate, along with these potassium rich substrates like calotropis, other weeds, tobacco residues, tomato crop residues, ash, poultry wastes are used. Over these 2-3 buckets of cattle dung slurry is sprinkled and covered with straw. This is finally covered with cattle dung and some old composted material or tank silt or ash, which acts as good casing material. The substrate is turned once in 15 days.

Windrow method

Turned windrows: The windrow method is a widely practiced method of composting in USA. In this process, the waste materials like garbage received are dumped and is sorted for plastic, glass and rubber. The waste materials are piled in long rows of 2 - 3 m width and 1.5 - 3 m height on a hard surface and usually in the open area (Plate 24). Aeration of windrow is by periodic turning using equipment such as front-end loader or specially designed machinery like aero tillers. Occasionally forced aeration in conjunction with turning has been applied to the windrow process. The windrow method is also exploited by distillery and sugar industry for composting of press mud with distillery-spent wash.

Plate 24 : Composting by windrow method

Passively aerated windrows: This method passively aerates by supplying air to composting materials through perforated pipes embedded in each windrow, thus, eliminates the need for turning. The plastic pipe is laid down in two rows, and 1.25 cm diameter holes are present in the pipe, the holes are oriented downward to minimize plugging and to allow condensate to drain. This method is commonly seen in Canada for composting seafood wastes with peat moss, manure slurries with peat moss, and solid manure with straw or wood shavings.

Factors Influencing Composting

There are numerous factors determining the fate of composting. It can be acceded that the time of composting and quality of the final product is affected by these factors. Hence, it is imperative to follow composting by employing optimized parameters; some of the grave important factors being:

C/N Ratio

The nitrogen and carbon content of the bio-mass substrate determine the rate of biodegradability of the substrate. The microbes in compost use carbon for energy and nitrogen for protein synthesis. Of the carbon present in the raw, uncomposted material, about 2/3 of it will be burnt away as it supplies energy to the microbes. This fraction is lost to the atmosphere as carbondioxide. The remaining 1/3 is incorporated into their cellular structure: amino acids, nucleic acids, lipids and polysaccharides. For every kilogram of usable carbon, approximately 320 grams of it goes directly into microbial cells. If analyzed, the average compost microbe cell would have a C:N ratio between 5:1 and 8:1. The ideal C/N ratio of 30:1 is suitable for microbial growth. The decomposition is slow in the environment, if the C/N ratio is greater than 60, since, microorganisms require 1 unit of nitrogen to degrade every 10 units of carbon. Most organic materials do not have this ratio and to accelerate the composting process, it may be necessary to balance the numbers. Hence, for composting of crop residues and leaf litter, which have a wider C/N ratio, the process can be hastened by mixing with blending agents like cattle dung, slurry or nitrogen rich plant materials. For optimization of C/N ratio for composting, proper blending material is required to hasten composting and to get good quality compost; approximation of ratio of C/N can be done using the formula

$$\text{C/N of the mixed substrate to be composted} = \frac{\text{Sum of C/N ratios of substrates used}}{\text{Number of substrates used}}$$

Example: If you have one bag of vegetable produce (C:N = 19:1) and one bag of barley straw (C:N = 100:1) then combined you have a C:N ratio of (19:1 + 100:1)/2= (119:1)/2 = 59.5:1

The C/N optimization is required, since, if there is wider C/N ratio, then the time for composting is increased and if there is narrow C/N ratio, then there is loss of valuable nitrogen to atmosphere in the form of ammonia. Addition of urea for enrichment, results in loss of nitrogen as ammonia volatilization. Solid wastes generally contain sufficient P for microbial growth and activity. The optimum carbon to phosphorus (C/P) ratio is around 100:1.

Now the concept of adjuvants is used. These are raw materials or waste products or byproducts or agro-residues or chemicals which provide easily assimilable nutrients to promote microbial growth, sustain their activity and complement in the composting process, by providing energy, thus, helping speedy conversion and maturation of compost. Urea and lime are examples of the common adjuvants used.

Moisture

For proper decomposition, the moisture content of the organic wastes has to be maintained at 60-70%; excess moisture brings in anaerobic condition and delays the process of decomposition. Moisture is essential to all living organisms, and most microorganisms, lacking mechanisms for moisture retention (like our skin) are particularly sensitive in this regard. At moisture content between 35-40%, decomposition rates are greatly reduced and below 30% they virtually stop. For most compost mixtures, moisture content of 55-60% is the recommended upper level.

Aeration

Composting is an aerobic process. Oxygen demand is very high during the initial decomposition stage that is characterized by a rapid rise of the microbial population, temperature and high rate of biochemical activity. Oxygen content is mainly maintained at a level above 10% whether it is done by turning or by forced aeration. Temperature raise is achieved only by the oxidation of the organic substances. Under anaerobic conditions many intermediary compounds formed such as volatile organic acids, volatile S and N compounds, cause foul odour. If oxygen is unavailable, the organic matter produces methane, hydrogen sulphide, skatoles, indoles, mercaptans etc., accounting for the characteristic strong odours. In horticulture, organic acids in such compost are toxic to the plants and remain phytotoxic over long periods.

There are three factors contributing to the maintenance of aerobic conditions within a composting material: bulk weight, porosity, and free air space. Moisture content is related to free air space and porosity. When void spaces within a compost pile are filled with water, the material has little strength and acts like a plastic. This stops aerobic composting due to the restriction of oxygen transfer. Only when the voids are filled with air, aerobic composting occurs. The free air space of compost can be increased with the addition of conditioning agents, such as recycled compost products, and bulking agents like wood chips. There is evidence that the minimum optimum moisture content for a given material is related to a certain minimum free air space percentage. Providing the optimum aeration is an important cardinal factor for obtaining high temperatures and for producing good quality compost. Frequent aeration in form of turning may lead to drying of the material; and, delayed turning may accumulate excess moisture, leading to anaerobic condition with foul smell development. Turning the compost material once in two weeks was found to be optimum for decomposing a mixture of wood waste and press mud blended in 1:1 proportion.

Particle size

Smaller the particle, faster is the decomposition rate as smaller materials have more surface area available for microbes to attack. Therefore, reducing the particle size of raw materials will increase the speed of the composting process. Size reduction also reduces the volume of the compost pile, thereby saving space. Particle size cannot be too small, for example, sawdust and wet materials can decrease aeration, reduce the rate of composting and perhaps cause anaerobic conditions leading to odour problems. Chopping of the substrate to 10 mm is recommended when windrows are formed with forced aeration.

Temperature

Temperature is one of the important parameter, as any change from the normal range is considered a faulty gesture of composting. Increase in temperature beyond optimum is not desirable as it is stressful for the survival of the microorganisms and reduces moisture and oxygen, which retards the rate of composting. This sets in partial decomposition of the composting mix, its partial desiccation and presumably death of certain mesophiles. The temperature indicates the decomposition rate of the substrate. Increase in temperature suggests rapid growth and activity of the microorganisms and this should be followed by optimum maintenance of the temperature for healthy compost. The harmonious temperature fluxes is very important for expediting composting; initial mesophilic temperature below 40°C followed by thermophilic activity from 40°C to 70°C, finally cooling phase with temperature dipping below 40°C have subsequent impact on the biodegradation.

Temperature generated within compost mass is important for two reasons: a) to maximize decomposition rates; b) to produce a material which is microbiologically safe for use. The increase in temperature is desirable as it eliminates most of the animal and plant pathogenic microorganisms, depending on the source of the substrate employed, and kills most of the weed seeds.

pH

The compost process will be more effective if the pH is in the range of 6.5 to 7.5. For initial mixtures in the range of 5.5 or 9.0, composting may take longer to establish. In particular, bacterial activity may be slow when pH is below 6.0. pH is likely to vary during composting process, but the final product will usually have a value near neutral.

Period of decomposition

The period of decomposition varies for the different types of organic wastes. For example, the urban solid wastes with high quantity of biodegradable substrate take 45 days to degrade. Coir pith composting takes about 150-180 days. Crop residues take about 90-120 days for composting. Pressmud takes about 60-90 days under optimum conditions of moisture, temperature and aeration with suitable microbial inoculation.

Rapid Composting Using Microorganisms to Hasten/Activate the Process

It generally takes about 4-6 months to get finished compost under normal conditions. Considerable research has gone to reduce the time of composting and to upgrade nutrient content. Rapid method will be more acceptable because there would be greater turnover in a lesser area of land with less effort. Fungi with ligno-cellulolytic activity such as *Asperillus niger, Penicillium* sp., *Trichoderma viride, T. ressei, Trichurus spiralis*, *Chaetomium* sp., *Phanerochaete chrysosporium*, *Pleurotus sajor-caju* and *Paecilomyces fusisporus* are used to hasten the decomposition process (Plate 25). These cultures are added at the rate of 2 kg per tonne. Several experiments have been conducted to study the role of efficient lignocellulolytic fungi in composting of crop residues richer in cellulose, hemicelluloses and lignin (Table 29).

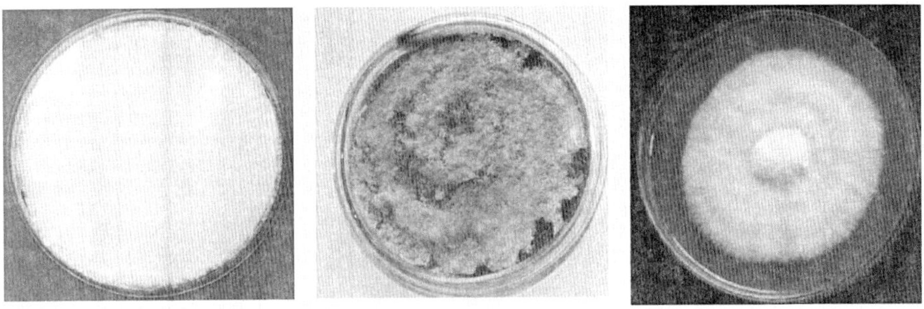

a b c

Plate 25 : Fungi used for rapid composting a. *Phanerochaete chrysosporium*, b. *Trichurus spiralis*, c. *Pleurotus sajor-caju*

Table 29: Enrichment and hastening of compost with lignocellulolytic fungi

Location	Substrates	Total N%		C/N ratio	
		without inoculum	with inoculum	without inoculum	with inoculum
New Delhi	Sorghum+ Wheat straw	1.38	1.82	20.6	12.0
New Delhi	Paddy straw	1.52	1.76	15.5	12.3
Hisar	Mixed crop residue	1.38	1.52	24.0	20.0
Pune	Sugarcane trash	0.98	1.54	35.0	24.0

Phanerochaete chrysosporium has been found to secrete efficient lignin degrading enzymes, selection of its over producer mutant has been suggested to facilitate rapid composting. Strategies have been developed for increasing the rate, efficiency and selectivity of the degradation of lignocelluloses by utilizing *Trichoderma viride* L-333, *Coriolus versicolor* 082 and *C. maxima* 0275. Many of the organic wastes are converted to compost by a consortium consisting of *Trichoderma viride* and *Pleurotus sajor-caju*.

Final Product

The physical and chemical properties of the final compost depend on the type of raw materials used *viz.,* agricultural, municipal and industrial wastes, along with the extent of decomposition by soil microorganisms. The compost may have a C/N ratio of 15-20 and organic matter content of 8-10%. Compost is considered superior if the C/N ratio is 20 or less and organic matter content is 30-60%. After a period of decomposition, the composted materials are removed from the yard, sieved in 4 mm sieves. The fine compost obtained, containing moisture of 30-40%, is stored in shade for further use.

Enrichment of Compost with Nutrient Enhancing / Biocontrol Microorganisms

Phosphocompost

India is producing 0.4% of world rock phosphate which is of inferior quality and not used for fertilizer production. Nearly 95% of P fertilizer in the country is from imported rock phosphate. India has 260 million tonnes of low grade rock phosphate, which can save a minimum of Rs.3,000 foreign exchange for every tonne of phosphatic fertilizer. The need for effective usage of low grade Indian rock phosphate as phosphatic fertilizer thorough composting has led to the process called phosphocompost. Rock phosphate can be added at the beginning of composting to get phosphorus enriched compost. The level of rock phosphate may be 12.5 - 25% depending on the required P enrichment. There is also a practice of addition of gypsum and some oil cakes, which are not used as cattle feed, during the last turning of the compost. Along with addition of rock phosphate, phosphate-solubilizing microorganisms like *Bacillus megaterium, Aspergillus awamori, Pseudomonas, Micrococcus, Flavobacterium, Penicillium, Fusarium* etc. are inoculated at the rate of 1-2 kg per tonne of the substrate. The phosphocompost is used as the phosphatic manure, which is superior to super phosphate as it serves as a good source of P to crop plants and is less chemically fixed in soils.

Addition of rock phosphate to wood waste-glyricidia mixture with subsequent inoculation with a phosphate solubilizer *Pseudomonas striata* resulted in a biocompost rich in phosphorus. *Pseudomonas striata* solubilized rock phosphate and released higher amounts of available P, hastening the composting process, besides increasing the P content in the biocompost.

Enrichment with plant growth promoting organisms

Nutrients status of the compost can be enhanced through inoculation with plant growth promoting organisms. Bacterial inoculants such as *Azotobacter, Azospirillum* are added to the decomposing materials at the rate of 2 kg / tonne after the thermophilic phase *i.e.* after a period of 30-45 days of initial decomposition. These nitrogen-fixing bacteria increase in population and enhance the nitrogen content. The fungal cultures such as *Trichoderma* spp. or bacteria like *Bacillus* spp. are also inoculated after 30-45 days initial decomposition, which on increasing in population improve the quality of the final product. These organisms can also help in suppressing root pathogens.

Suppressive compost

Soil-borne diseases result from a reduction in the biodiversity of soil organisms, which can suppress pathogens. Restoring important beneficial organisms that attack, repel, or otherwise antagonize disease causing soil organisms will reduce their populations to a manageable level. Beneficial organisms can be added directly, or the soil environment can be made more favourable for them with compost and other organic amendments. There are different phases in the production of compost, as said earlier.

The first phase occurs during the first few days when temperatures rise and sugars and other easily biodegradable substances are consumed. Over the next several weeks, in the second phase, the temperature range increases and cellulose and other less biodegradable substances, pathogens, and some biocontrol agents are destroyed. In the third phase, temperatures and decomposition rates decline as the supply of readily biodegradable substances becomes limited. The drop in temperature allows for microorganisms (e.g. *Bacillus, Enterobacter, Flavobacterium, Pseudomonas,* and *Streptomyces*) to recolonize the compost's interior layers, thereby creating the natural suppressiveness.

There are two primary mechanisms by which the colonies of biocontrol organisms in compost combat disease: general suppression and specific suppression. General suppression occurs when a high-microbial activity environment is created in which the germination of pathogen propagules is inhibited. Specific suppression involves the action of a specific microbial agent in suppressing a specific pathogen. This can be achieved by inoculating the compost with the desired microbial agent. Inoculation of mature composts with biocontrol agents has improved the consistency as well as the spectrum of disease suppression, particularly for *Rhizoctonia solani*. Isolates of several *Trichoderma* spp. can provide effective control of this disease, and even better control if applied in combination with any of several bacterial biocontrol agents.

Compost enriched with the antagonistic fungus *Gliocladium virens*, and the antagonistic bacteria *Flavobacterium balustinum*, *Pseudomonas putida* and *Xanthomonas maltophilia* have also been used as effective suppressive compost controlling diseases caused by *Phytophthora cinnamomi*, *Pythium ultimum*, *Rhizoctonia solani* and *Fusarium oxysporum*. In the nurseries, biocontrol agent-fortified composts seem to offer the greatest opportunities. For example, compost fortified with *Flavobacterium balustinum* and *Trichoderma hamatum* has been very effective for control of *Fusarium* wilt, *Rhizoctonia* rot, in addition to *Phytophthora* wilt and *Pythium* damping-off diseases. Further it reduced the cost of nursery production by nearly 50% when used instead of drenching the potting mix with fungicides. Suppressive compost must be popularized amidst farming community, as it is a simple method to control soil-borne pathogens important in agriculture, horticulture and forestry.

Stability and Maturity of the Compost

The ultimate goal in composting is to produce a humus-like product that can be used for soil improvement and plant growth, and to reduce environmental pollution. The compost, thus, prepared must be allowed to stabilize and mature. Stability is a stage in the decomposition of organic matter and represents the state of microbial activity. Maturity is an organochemical condition of the compost, which indicates the presence or lack of phytotoxic organic acids. Thus, the two terms "stability" and "maturity" are not synonymous, though often used interchangeably. As stability represents the state of microbial activity, measurement of respiration through CO_2 evolution is often done. If the compost is stabilized it will have less microbial activity (as it would have utilized all degradable components in the substrate) and, thus, less CO_2 evolved through respiration.

Immature composts induce high microbial activity in soil potentially causing oxygen deficiency in soil and biological blockage of available soil nitrogen. They also leave odouriferous phytotoxic compounds into the soil *viz.*, ammonia, ethylene oxide and low molecular weight fatty acids. Plant health and disease suppressive properties of compost are affected by maturity, hence, the compost maturity and stability is of great importance. The recommended compost stability index is given in table 30.

Table 30: Compost stability index based on carbon dioxide evolution

Respiration rate (mg CO_2-C/g compost C-day)	Rating	Characteristics
< 2	Very stable	Well cured; no malodours; earthy odour
2-5	Stable	Cured compost; minimal impact on soil dynamics
5-10	Moderately stable	Uncured compost; some malodour; addition to soil may immobilize N; high phytotoxicity potential; not recommended for growing plants from seed
10-20	Unstable compost	Very immature compost; high malodour and phytotoxiciy potential; not recommended for growing plant from seed
>20	Very unstable compost	Extremely unstable material; very high malodour and phytotoxicity potential; not recommended for use.

Alternatively higher levels of ammonical N in the compost indicates that it is not cured, whereas presence of nitrate N indicates it is properly cured (ripened).

A plant assay is used to evaluate maturity of the compost. The presence of phytotoxic compounds indicates that the compost is not matured. Seed germination in compost is done to test the compost for maturity.

Procedure

(i) Fill the pots with soil, which serves as control.

(ii) Fill the pots with compost, which serves as compost treatment.

(iii) Sow 10 seeds in each of soil and compost pots. Keep 6 replications of soil and compost treatments.

(iv) Water the pots regularly to keep the pots moist.

(v) After a week, count and record number of seeds germinated in each container and calculate germination rate.

$$\text{Average percentage germination} = \frac{\text{Average number of seeds germinated}}{10} \times 100$$

If the germination rate in the compost pots is less than the germination rate in soil control pots, then the compost is considered to be immature and needs further curing. The general compost quality standard is given in table 31.

Table 31: Standards for evaluating the quality of compost

Parameters	Good quality	Poor quality
Colour	Brown to black	Varies
Smell	Pleasant earthy smell	Bad odour
pH	6.5-8.0	< 6, > 8
C/N ratio	10-20	< 10, > 20
Temperature	30-45°C	> 45
Moisture	25-30%	>30
Humus percentage	> 4%	< 4%
Plant assay	Good growth	Growth inhibition
Nitrogen	> 1.25%	< 1%

Some of the common problems encountered during composting and its remedies are given in table 32.

Table 32: Some of the common problems faced during composting and its remedies

Problem	Cause/Clues	Solution
Nothing rots	Not enough moisture	Add enough moisture
	Material too woody and not enough nitrogen	Shred the woody substrate till optimum particle size
	Pile too small	Add nitrogen rich materials
Rotten and Sulfurous odour	Too many food scraps or food wastes	Avoid food wastes, or mix food wastes evenly throughout the pile
	Too many grass clippings	Mix grass clippings with carbon rich dry substrate
	Anaerobic conditions	Open pile apart, rebuild the pile loosely with different materials
	Very small particles	Turn pile while adding material of different sizes and coarseness. Layer pile with plenty of loose and bulky material
	Too wet	Cover the pile durin g monsoon, turn the pile while adding some dry, high carbon bulking agent, which absorbs excess moisture
	Pile too large	Decrease the pile size
Ammonia smell, mostly when turning the pile	High nitrogen level	Add high-carbon amendments
	High pH	Lower pH with acidic ingredients and / or avoid alkaline ingredients
	Slowly available carbon source, like large woody particles; C:N ratio less than 30:1	Use another carbon amendment or increase the carbon proportion
	Anaerobic condition	Turning the pile helps circulating oxygen
Site related odours (piles not odourous)	Raw materials, odour is characteristic of the raw material	Handle raw materials promptly with minimal storage
	Nutrient rich puddles because of poor drainage, standing puddles of water	Divert runoff away
Pile has low Temperature	Gradually falling temperatures; pile does not reheat after turning or aeration as composting nearing completion	None required

	Low moisture, cannot squeeze water from materials	Add water and remix
	Pile too small, or if the surroundings are cool	Rebuild the pile with optimized parameters, if cold surroundings increase the pile size
	High C/N	Add sufficient nitrogen rich materials
	Too dry or too wet	If too moist, the pile will have an odour, then rebuild the pile and add dry carbonaceous material. Cover when raining. A dry pile having no odour, rebuild the pile while wetting the materials, a cover may be needed to retain the moisture
	Material too coarse or too fine	Add materials having sufficient particle size to counteract the effect of small or coarse sized particles
	Anaerobic conditions	Aerate the pile
	pH low, pH measures less than 5.5; garbage-like odour	Add lime or wood ash and remix
Pile too hot	Insufficient aeration for heat removal, pile is moist	Turn pile to increase airflow
	Moderate to low moisture; limited evaporative cooling, pile feels damp but not excessively wet or dry	Add water; continue turning and aeration to control temperature
	Pile is too large	Decrease the pile size to optimum
	Too much nitrogen	Turn pile while incorporating more carbon-rich materials
	Pyrolysis or spontaneous combustion, low moisture content; pile interior looks or smells charred	Decrease pile size; maintain proper moisture content; add water to charred or smoldering sections; break down pile, combine with other piles
	Compost is not stable, High temperatures or odours in curing or storage pile	Manage pile for temperature and odour control, turn piles as necessary; limit pile size
Uneven temperatures or varying odours in pile	Materials at different stages of maturity, temperature varies along the pile length	None required
	Poorly mixed materials, visible differences in the pile moisture and materials	Turn or remix pile
	Uneven airflow or air short-circuiting, visible differences in the pile moisture and materials	Remix pile

Compost contains clumps of materials and large particles; texture is not uniform	Poor mixing of materials or insufficient turning, original raw materials discernible in compost	Screen compost; improve initial mixing
	Airflow uneven or short-circuiting, Wet clumps of compost	Screen or shred compost; improve air distribution
	Raw materials contain large particles and non-degradable or slowly degradable materials	Screen compost; grind and / or sort raw materials
	Active composting not complete	Lengthen composting time or improve compost conditions
Flies and insects in and around the pile	Flies breed in compost piles, fresh manure or food material at pile surface; flies hover around piles	Turn piles every four to seven days; cover static piles with a 6-inch layer of compost, or add lime and soil to top of the compost material and turn it the next day
	Flies breeding in raw materials, wet raw materials stored on site more than four days	Handle raw materials promptly
	Too much food wastes	Add the food wastes only to the middle of the pile, if problem persists discontinue the use of food wastes
Mice and rats in pile	Food wastes	Try avoiding meat, dairy, animal faeces, vegetable and fish wastes
	Presence of burrows	Aerate the pile as per the optimized parameters

In conclusion it can be said that

(i) Compost can supply all the 14 essential nutrients for plant growth obtained from soil.

(ii) Compost acts as a soil conditioner as it improves physical condition of the soil such as soil aggregation, soil structure, water holding capacity and cation exchange capacity. Thus, it provides a plethora of goodies like efficient seed germination, easy root penetration, adequate root aeration etc.

(iii) It is an effective material providing resistance of soil to slacking, surface sealing, wind erosion etc.

(iv) It improves biological activity of soil, especially the enriched compost enhances the numbers of plant growth promoting organisms and those biologically controlling root pathogens.

(v) Most of the human and animal pathogens and weeds are killed during the thermophilic phase.

(vi) It reduces the application of chemical fertilizers deleterious to the environment.

(vii) It has better residual effect, hence helps subsequent crops.

(viii) Humic and fulvic acids present in the compost play an important role in promoting plant growth.

(ix) It helps to increase the efficiency of minerals and biofertilizers.

(x) More than anything it mitigates environmental pollution by effluents coming out of industries. Thus, it is a boon to industries to transform effluents, which is considered to be a pollutant to eco-friendly product, which helps the farming community.

References

Acharya, C.N. 1939. Comparison of different methods of composting waste materials. *Indian Journal of Agricultural Science*. 4: 135-149.

Bagyaraj, D.J. and Radhakrishna, D. 2000. Recent advances in composting. In: *Vistas in Mycology and Plant Pathology*. (eds. L.V. Gangawane, Y. Ganju, S. Thomas and S.L. Aggarwal). Commonwealth Pub., New Delhi, India. pp. 88-100.

Bagyaraj, D.J., Sharath, S. and Balakrishna, A.N. 2006. Composting Technology: Significance and Potential. In: *Current Concepts in Botany*. (eds. K.G. Mukerji, and C. Manoharachary). I.K.International Pub., New Delhi, India. pp.287-320.

Balakrishna, A.N., Srinivas, K.N., Radhakrishna, D. and Gowda, T.K.S. 1991. Enrichment of municipal waste compost (MPC) with urea. *Current Research*. 20: 160-162. Bhardwaj, K.K. and Gaur, A.C. 1985. *Recycling of Organic Wastes*. ICAR Pub., New Delhi, India.

Chongrak Polprasert, 1996. *Organic Waste Recycling* (2nd edn). John Wiley & Sons, Sussex, England.

Davis, C.L., Hinch, S.A., Donkin, C.J. and Germishuizen, P.J. 1992. Changes in microbial population numbers during the composting of pine bark. *Bioresearch Technology*. 39:85-92.

Gaur, A.C. 1999. *Microbial Technology for composting of Agricultural Residues by Improved Methods*. ICAR Pub., New Delhi. India.

Hagerty, D.J., Pavoni, J.I. and Jr. Heer, J.E. 1973. *Solid Waste Management*. Von Nostrand Reinhold, New York. U.S.A.

Hajra, J.N. and Kole, S.C. 1998. Enrichment of compost with inoculants. *Biofertilizer Newsletter*. pp. 2-4.

Murkute, S.B., Thopate, A.M., More, B.B. and Hapase, D.G. 1992. Biodegradation of agro industrial wastes. Proceedings of National Seminar on Organic Farming. pp. 67-68.

Radhakrishna, D. and Bagyaraj, D.J. 1998. Microbial degradation of solid waste as a source of plant nutrients. In: *New Trends in Microbial Ecology*. (eds. B. Rai and I.S. Dkhar) N.E.Hill University Pub., Shillong. India. pp. 108-117.

Ravindran, A.D.2005. *Fundamentals of Composting*. +G Publishing, Dindugal, India.

Robinson, C.W. and Howell, J.A. (eds.). 1985. *Composting: Comprehensive Biotechnology*. Pergamon Press, Oxford, UK.

Sharath, S. and Jagadeesh, K.S. 2004. Field scale composting of wood waste generated in a paper mill. *Bulletin of Agricultural Science*. 1: 9-18.

World Bank Technical Report on Bark and Sawdust Composting. 1995. Vol. 264: 63-65.

Zucconi, F., Pera, A., Forte, A. and De Bertoldi, M. 1981. Evaluating toxicity of immature compost. *Biocycle*. 22: 54-57.

Chapter-10

Mass Production and Quality of Microbial Inoculants

Microbial inoculants in agriculture, horticulture and forestry can be defined as preparations containing strains of microorganisms which can augment the microbiological process such as nitrogen fixation, phosphate mobilization, excretion of growth promoting substances, decomposition of organic wastes, biocontrol of insect pests and diseases etc. Thus they become an important component of sustainable agriculture. Microbial inoculants are based on renewable sources available in soil and nature and are low cost input and eco-friendly. These inoculants carrying the microorganisms when applied to soil/ plant promote plant growth and productivity.

Microbial inoculants aiding plant nutrient supplies, decomposition of organic wastes and control of plant diseases and pests are discussed in this chapter. Microorganisms supplying plant nutrients like nitrogen fixers and phosphate solubilizers, decomposing organic wastes and controlling insect pests and plant diseases have been discussed in detail in the previous chapters. The identified microorganisms responsible have to be mass produced so that it is available to the end user, the farmer. Mass production technology developed for different microbial inoculants are narrated below. In India, Bureau of Indian Standards (BIS) (formerly ISI) listed the Indian standard specifications for microbial inoculants earlier. The present specifications for quality control of biofertilizer organisms are given through Fertilizer Control Order (FCO). For biocontrol organisms it is given by Central Insecticide Board Regional Committee (CIBRC).

Rhizobium

Mass culture production of rhizobia had its beginning in 1895 when Nobbe and Hiltner introduced a laboratory grown culture of *Rhizobium* sp. under the name 'Nitragin'. Prior to 1923 cultures of rhizobia were mostly prepared in small bottles or cans and in 1927 a method for large scale production of rhizobia by submerged fermentation with the development of a device for artificial aeration was developed. There are several forms of inoculants *viz.* agar-based, broth cultures, frozen concentrate, carrier based-granular soil inoculum, porous gypsum granules, natural peat granules and freeze-dried which can be used for inoculating seed or soil. The carrier based inoculant is the most acceptable form which are produced today since the use in this form results in better survival of rhizobia.

The first requirement for the mass production of a microbial inoculant is that it should have certain desirable characteristic like

(i) Ability to multiply under broth culture conditions;

(ii) Survival in carrier material;

(iii) Stability during storage;

(iv) Ability to withstand adverse conditions (environmental stress, agrochemical inputs) and to perform desired action under natural/field conditions;

(v) Competitiveness in terms of enhanced/early colonization of soil/rhizosphere and internal plant organs as per the mode of action; and

(vi) Enhanced establishment in a wide range of host genotypes.

Production of *Rhizobium* inoculant

Having selected a suitable bacterial strain that can form effective nodules in association with selected legumes over suitable range of environmental conditions, it is necessary to lyophilize them to maintain their efficiency, since sub-culturing at monthly intervals on artificial medium may lead to deterioration of the activity of the organism.

Whenever it is required for use, the tube containing the lyophilized material is cut open and the lyophilized cells are reconditioned. These are grown in sufficient numbers on agar slants to be used for mass multiplication during the season. It is to be recognized that the rhizobia are sufficiently mutable to require their regular continued evaluation with the relevant hosts. Such a programme should be continued with the systematic long-stored, untouched culture in lyophilized conditions as mentioned above. This is necessary to ensure continued symbiotic efficiency of the cultures used as seed inoculum.

Mass-scale propagation of rhizobia can be carried out using system of rotary shaker or fermenter. About 100 ml aliquots of the yeast extract mannitol broth (YEM) are dispensed into 250 ml conical flasks, which are plugged with non-absorbent cotton and are sterilized at 15 lb pressure for 20 minutes and cooled after sterilization. Three- to six-day-old cultures of *Rhizobium* are examined for the purity and about 4 to 6 ml of sterile broth is transferred aseptically into the tubes. The bacterial growth is scrapped with the help of an inoculation needle and the resulting bacterial suspension is transferred to a 250 ml conical flask containing YEM broth. The flasks are incubated at 30°C on a rotary shaker for 2 to 7 days depending on the type of *Rhizobium*. Fermenters are used for industrial scale production of biofertilizers. Culture vessels ranging from 5 to 1,000 litres can be used depending upon the requirement. The amount of inoculum culture to be added into the fermenter vessel depends on the size of the fermenters, but the ratio between the inoculum and the medium in the vessel should be maintained at 1:20 (5% inoculum rate). The broth is continuously aerated by forcing sterile air through porous stainless steel sparger. Various fermentation requirements like aeration, agitation and fermentation time vary from strain to strain. The process optimization for some of the rhizobial cultures using fermenter-based technology is presented in table 33.

Table 33 : Optimum fermentation conditions for mass multiplication of *Rhizobium* strains.

1. Type of reactor	Stirred tank
2. Type of operation	Batch
3. Carbon source	Sucrose or molasses (3 - 5 g/l)
4. Nitrogen source	Corn steep liquor or Yeast extract
5. pH	7.8 (controlled)
6. Temperature	28°C
7. Inoculum rate	10% (v/v)
8. Inoculum count	10^9 cells/ml
9. Antifoam	LPG
10. Aeration	0.5 VVM (volume of air/ vol. of liquid/min)
11. Agitation	Depends on the fermenter size
12. Fermentation time	Variable (18 - 40 hrs)
13. Cell processing	Centrifugation or membrane filtration followed by lyophilization
14. Quality specification	10^8 /g carrier at the time of manufacture
15. Shelf life	6 months from date of manufacture

When the number of rhizobia in the broth has attained the required standard (10^8 -10^9 cells/ml), the broth should be added to the carrier for preparation of carrier based inoculant.

Carriers for rhizobial inoculants

The medium in which rhizobia are allowed to multiply is an important factor in rhizobial culture preparation. The term 'carrier' is generally used for a medium which carries the live microorganisms. As per the quality control specification, the carrier should be in powder form and capable of passing through 150 - 212 micron (72 - 100 mesh) IS sieve. A good carrier material should:

(i) have high water holding capacity;
(ii) be non-toxic to rhizobia;
(iii) be easy to sterilize by autoclaving or gamma irradiation;
(iv) be readily and inexpensively available;
(v) provide good adhesion to seed;
(vi) have pH buffering capacity; and
(vii) have cations and/or anions exchange capacity

Different carrier materials like peat, lignite, charcoal, rice husk, pressmud, flyash, vermiculite, soil and coir dust have been employed. Lignite and charcoal are more commonly used as carrier in India.

Preparation of carriers

The carrier material is sun-dried to 5% moisture and pulverized to fineness of 72 - 100 mesh. Coarser materials form balls on mixing with broth and seeds are not properly coated with the inoculant. Lignite powder being acidic (pH 5.5) usually 1 kg of lime ($CaCO_3$) is mixed with 9 kg of lignite to bring the pH to about 7. Charcoal powder being neutral in pH does not require lime. To enhance its effectivity for multiplication of rhizobia, 1% $CaCO_3$, 0.5% KH_2PO_4 and 25% soil are thoroughly mixed with charcoal. Finally the carrier is mixed with 10% water before sterilization. The carrier material must be sterilized before blending with broth culture to suppress the proliferation of contaminants. Better growth of *Rhizobium* is obtained in sterile carrier. Best method is γ irradiation (5 megarods) for carrier sterilization. The carrier materials may also be packaged and sealed in autoclavable high density polypropylene bags which can stand a temperature of 121°C and sterilized in an autoclave.

Blending of broth culture with carrier

For preparation of carrier based inoculants, the rhizobial number of broth must attain a minimal population of 10^9 cells/ml which is mixed with sterilized carrier. The broth is added to one-third of the water-holding capacity of carrier. Blending is done by thorough mixing of broth culture with sterilized carrier in trays manually. In bulk operation, mechanical mixers may be used. After blending the moisture in packet would range from 40-60%.

Packing and labeling

The blended carrier is kept for 2-4 hours for curing. After curing, the inoculant is packed in low density polypropylene bags of 0.2 to 0.5mm gauge size either manually or by using automatic dispenser. The bags are sealed by an electric sealer leaving about two-third vacant space to give proper aeration to the inoculant. Low density polypropylene (LDPE) packets are used universally, as they can be readily sealed, allow high gas exchange with relative low moisture loss. The inoculant packet is labeled with relevant information regarding culture *viz.* name of the culture, crop for which intended, expiry date, directions for storage / use and precautions, certification mark if any. Packets (Inoculants) must be inoculated for a week in the room at 30° C so that the bacteria multiplies and attains the required count. The packet may then be stored in a cold room (4-15°C).

The inoculant should not be subjected to excessive temperatures, ideally it should be held at 4°C throughout transport and storage. Inoculum in bulk should be transported in strong cartons or boxes. Distribution of cultures during summer and rainy season requires suitable arrangements and strict quality control measures. The mass multiplication scheme is shown in fig. 22. The shelf life of *Rhizobium* inoculant thus prepared is about 6 months at room temperature and in a cold room it can be up to 4 years. During the end of this period each gram of the inoculant shall contain 10^7 cells/g of the carrier on dry weight basis.

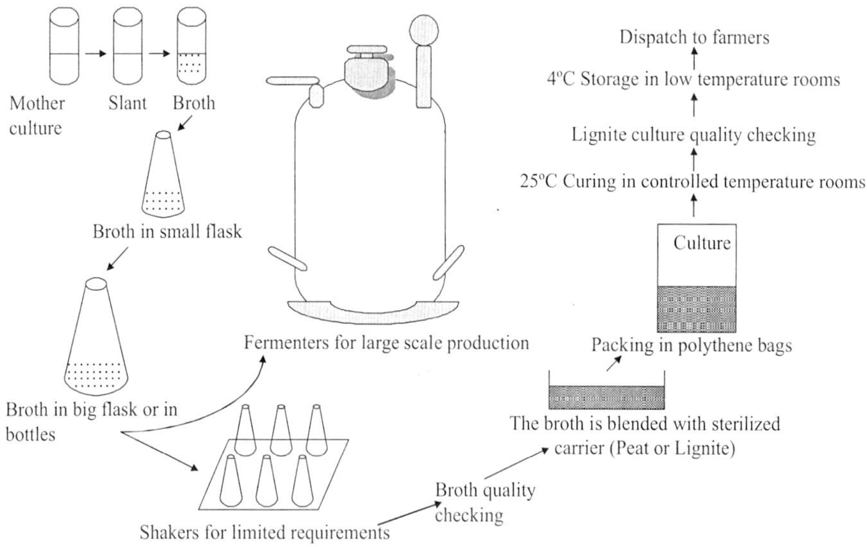

Fig. 13 : Method of mass multiplication of rhizobia

Crops on which they are to be used

Rhizobium biofertilizers can be used for cultivating legumes like, redgram, bengalgram, cowpea, greengram, blackgram, horsegram, soybean, beans, clusterbeans, groundnut etc. Care should be taken to use *Rhizobium* only to those crops for which they are meant as mentioned on the biofertilizer packet, as some of them are host specific. Usually, 400-500 g, lignite or charcoal-based culture would be sufficient to the quantity of seed required per hectare. The method of seed inoculation includes preparation of 10% sugar or pharmaceutical grade gum arabic or 1% carboxy methyl cellulose (CMC) solution. This solution is sprinkled on the seeds and the seeds are thoroughly mixed so as to have a uniform coating. A count of 1,000 viable cells per seed is to be attained at the time of treating the seed and quantity of culture used is accordingly adjusted. The seeds are spread uniformly for drying on a gunny bag or cement floor in shade avoiding direct sunlight. The seeds are sown immediately (preferably in the early morning or late evening).

Certain precautions are to be observed in the use of rhizobial cultures. The culture has to be used before the expiry date marked on the packet and the inoculated seeds are not allowed to come in direct contact with any pesticides or fertilizers. If fungicides, insecticides, herbicides and chemical fertilizers have to be used, there should be a gap of atleast 15 days between the use of these chemicals and biofertilizer. In order to protect rhizobia from effects of fertilizers, pesticides, acid / saline and dry soils, pelleting of the seed is the most commonly used practice. The most commonly used pelleting agents are calcium carbonate, rock phosphate, charcoal, talc, gypsum, bentonite etc. Co-inoculation with other biofertilizers like phosphate solubilizing microorganisms, AM fungi etc. have no deleterious effect on *Rhizobium* inoculant.

Quality control of rhizobial inoculants

The quality of rhizobial inoculants is of great importance in ensuring field performance as well as for the commercial prospects of inoculant industry. These specifications are given below.

Quality control specifications prescribed for *Rhizobium*

(i)	Base	Carrier based* in form of moist/dry powder or granules, or liquid based
(ii)	Viable cell count	cfu minimum 5×10^7 cells/g of powder, granules or carrier material or 1×10^8 cells/ml of liquid.
(iii)	Contamination level	No contamination at 10^5 dilution.
(iv)	pH	6.5-7.5
(v)	Particle size in case of carrier based material	All material shall pass through 0.15-0.212 mm IS sieve.
(vi)	Moisture percent by weight, maximum in case of carrier based	30-40%
(vii)	Efficiency character	Should show effective nodulation on all the species listed on the packet

* Type of carrier

The carrier material such as peat, lignite, peat-soil, humus, wood charcoal or similar material favouring growth of the organism.

The inoculant packets should have the following information:

a) Name of the product
b) Name and address of the manufacturer
c) Name of carrier
d) Batch number
e) Date of manufacture
f) Date of expiry
g) Net mass
h) Crops for which intended
i) Storage instructions

Some Do's and Dont's are suggested for entrepreneurs, dealers and farmers dealing with *Rhizobium* inoculants.

Do's and Dont's for entrepreneurs, dealers and farmers

Do	Don't
❒ Keep biofertilizer packets away from direct heat and sunlight. Store it in cool and dry place	❒ Don't store biofertilizer packets under heat and sunlight
❒ Sell only biofertilizer packets which contain batch number, the name of the crop, the date of manufacture and expiry date.	❒ Don't sell biofertilizer packets after their expiry period is over
❒ If the expiry period of the packet is over, discard it. It will not be effective.	❒ Don't prick holes into the packets or puncture them
❒ Keep biofertilizer packets away from fertilizer or pesticide containers	❒ Don't store heavy items above biofertilizer packets
	❒ Do not mix the biofertilizer with fungicides, insecticides, herbicides and chemical fertilizers.

Liquid Biofertilizers

Recently, there is some interest to produce liquid biofertilizers. First of all it should be understood that a liquid biofertilizer is not the usual broth culture from the fermenter or a water suspension of the carrier based biofertilizer, as often mistaken. They are special liquid formulation containing not only the desired microorganisms and their nutrients but also special cell-protectents or chemicals that promote formation of resting spores or cysts for longer shelf-life and tolerance to adverse conditions.

To produce a high quality liquid biofertilizer the following things are essential.

- A suitable fermentation medium to support the growth of inoculant microorganisms cfu per ml, which contain cell protectents and chemicals promoting the formation of resting cells.
- A suitable pressure vessel fermenter which facilitates aseptic culture of desired microorganism and provides optimum growth conditions like temperature, aeration etc.
- A proper mechanism for aseptic dispensing of liquid culture into containers.
- Aseptic containers, which allow slow exchange of gas and are leak proof.

The advantages of liquid biofertilizers are

(i) High cfu can be maintained while treating seeds prior to sowing.

(ii) Contaminants can be easily identified and discarded.

(iii) A longer shelf life, make them more attractive commercial proposition.

(iv) Quick test for quality at production centre and on farm level is possible.

(v) They are cost concern input when compared to carrier based but have good potential of improving productivity by using lesser quantity.

(vi) They have greater potential to fight with native microbes and maintain soil health.

The first report on liquid inoculants for seed inoculation of *Rhizobium* was published from Holland in 1953. From then onwards liquid formulations have been studied by various workers not only for rhizobia but also for different biofertilizer organisms. These workers reported that liquid formulations of bradyrhizobial strain recorded higher acetylene reduction activity and better yield of soybean when compared with lignite carrier based formulation. Hence with high cell count, zero contamination, longer shelf life, greater protection against environmental stresses and convenience of handling are the some of the advantages of liquid inoculants over the conventional carrier based inoculants.

The viability of cells in a liquid biofertilizer is maintained for a longer period by addition of certain cell protectants and chemical amendments like glycerol, polyvinylpyrrolidone (PVP), trehalose, Fe-EDTA etc. Glycerol acts as a C source, PVP has a high water binding capacity and appears to slow down drying of inoculum after application. Trehalose is a very sensitive and responsive compound to stress conditions and also functions as a membrane stabilizer and protectent.

The liquid biofertilizer of good quality holds great promise for the following benefits over the conventional carrier based biofertilizers like longer shelf-life, saving on carrier material, transport, pulverization and sterilization, convenience in handling, storage and transportation and better performance. The use of liquid biofertilizer is just taking off from research and development stage and has great potential in future.

Azotobacter

Azotobacter is a well-known free-living nitrogen fixing bacterium. It is used as a biofertilizer for all non-leguminous crops especially in cereals like rice, maize, wheat etc. and in crops like cotton, jute, tobacco, vegetables, ornamental plants and plantation crops. They are known to secrete various growth promoting substances like indole acetic acid, gibberlic acid, cytokinins and vitamins which promote seed germination and induce vigour to the plant to get luxuriant growth. *Azotobacter* responds well to organic matter addition and exhibits better performance in presence of either compost or farmyard manure. This organism also has the ability to inhibit the growth of few soil-borne fungal pathogens.

Procedure for mass production is more or less similar to that of *Rhizobium* production except the media used and a few other differences. The media commonly used for production are Ashby's, Jenson's or Waksman no. 77 medium.

Crops on which they are used

It can be used for a variety of crops like cereals, mulberry, vegetables, fruit crops, plantation crops, flowering ornamentals etc. They are not host specific.

Dosage, method and time of application

Seed treatment: Seed treatment is done at the rate 400 g/ha. Method of seed treatment is same as that given under *Rhizobium* biofertilizer.

Soil application: 10 kg/ha mixed in 20 kg FYM and can be applied by broadcasting or by row method.

Slurry treatment / root dip method: A thick slurry of 1kg inoculum in 5 litre water is made and roots of seedlings are dipped in slurry for a minimum period of 30 minutes and then planted in the main field.

For crops more than 6 months duration *Azotobacter* biofertilizer can be applied to soil once in 6 months.

Shelf life

Biofertilizer packets normally have a shelf life of 6 months at room temperature. During the end of this period each gram of the inoculant shall contain 10^7 cells/g of the carrier (on dry weight basis).

Compatibility : Similar to *Rhizobium* biofertilizer

Quality control specifications prescribed for *Azotobacter* biofertilizer

(i)	Base	Carrier based* in form of moist/dry powder or granules, or liquid based
(ii)	Viable cell count	cfu minimum 5 x 10^7 cells/g of carrier material or 1 x 10^8 cells/ml of liquid.
(iii)	Contamination level	No contamination at 10^5 dilution.
(iv)	pH	6.5-7.5
(v)	Particle size in case of carrier based material	All material shall pass through 0.15-0.212 mm IS sieve.
(vi)	Moisture percent by weight, maximum	30-40%
(vii)	Efficiency character	The strain should be capable of fixing at least 10 mg of nitrogen/g of sucrose consumed

* Type of carrier

The carrier material such as peat, lignite, peat-soil, humus, wood charcoal or similar material favouring growth of the organism.

Marking on packets should be similar to *Rhizobium* biofertilizer.

Do's and Don'ts for entrepreneurs, dealers and farmers

Similar to *Rhizobium* biofertilizer.

Azospirillum

Azospirillum is a comma-shaped bacterium, containing large amount of lipid granules inside its cell. It enters the cortical cells of the root and fixes atmospheric nitrogen. Since these bacteria enter the root cells and fix atmospheric nitrogen without forming root nodules, they are said to be in associative symbiosis with the host. They also produce plant growth promoting substances.

Procedure for mass production

Desired *Azospirillum* obtained from an authentic source is grown on Okon's/Dobereiner's broth in small (250ml) conical flask.

(i) Inoculum from small flask is transferred to large flasks (1-2 litre capacity) and placed on a mechanical shaker for 5-7 days.

(ii) Under optimum conditions the growth of *Azospirillum* attain a population of 10^8 or 10^9 cells / ml in 5-7 days.

(iii) The culture obtained in flasks is called as starter culture and can be used for large scale production.

(iv) Large quantity of liquid medium (broth) is prepared and is placed in large flasks or bottle or fermenter. Inoculum from the starter culture at the rate of 1% by volume is added and the culture is allowed to grow for 5 to 7 days.

(v) Pellicle formation 1 to 2 mm below the upper surface of the medium is a clear indication of the growth of *Azospirillum.*

(vi) The culture is then mixed with a carrier material and packed in polythene bags.

(vii) Lignite, charcoal and vermiculite are the common carrier materials.

Crops on which they are used

It can be used for number of crops like paddy, ragi, maize, sorghum, fodder crops, tuber crops etc.

Dosage, method and time of application

Seed treatment: Seed treatment is done at the rate 400 g/ha. Method of seed treatment is same as that given under *Rhizobium* fertilizer.

Soil application: 10 kg/ha mixed in 20 kg FYM and can be applied by broad casting or by row method.

Slurry treatment/Root dip method: A thick slurry of 1 kg inoculum in 5 litre water is made and roots of seedlings are dipped in slurry for a minimum period of 30 minutes and then planted in the main field.

For crops with more than 6 months duration, *Azospirillum* biofertilizer can be applied to soil once in 6 months.

Shelf life

Biofertilizer packets normally have a shelf life of six months at room temperature. During the end of this period each gram of the inoculant shall contain 10^7 cells/gram of the carrier (on dry weight basis).

Compatibility: Similar to *Rhizobium* biofertilizer

Quality control specifications prescribed for *Azospirillum* biofertilizer

(i)	Base	Carrier based* in form of moist/ dry powder or granules, or liquid based
(ii)	Viable cell count	cfu minimum 5 x 10^7 cells/g of powder/ granules or carrier material or 1 x 10^8 cells/ml of liquid.
(iii)	Contamination level	No contamination at 10^5 dilution.
(iv)	pH	6.5-7.5
(v)	Particle size in case of carrier based material	All material shall pass through 0.15-0.212 mm IS sieve.
(vi)	Moisture percent by weight, maximum in case of carrier based	30-40%
(vii)	Efficiency character	Formation of white pellicle in semisolid nitrogen-free bromothymol blue media

* Type of carrier

The carrier material such as peat, lignite, peat-soil, humus, wood charcoal or similar material favouring growth of the organism.

Marking on packets should be similar to *Rhizobium* biofertilizer.

Do's and Don'ts for entrepreneurs, dealers and farmers

Similar to *Rhizobium* biofertilizer

Phosphate Solubilizing Microorganisms

Plants generally take up phosphorus as primary orthophosphate ion (H_2PO_4) and sometimes as secondary orthophosphate ion (HPO_4^{-2}). Phosphorus nutrition in the early stages benefits the plant by producing deeper and abundant roots. Tropical soils are mostly deficient in available phosphorus, though the soil may contain P abundantly in unavailable form (it is a form plant roots cannot take). Several bacteria, fungi, actinomycetes and algae have the capacity to solubilize insoluble mineral phosphates and have been reported to solubilize varying quantities of phosphorus depending on the efficiency of the strains. These groups of microorganisms are collectively called phosphate solubilizing microorganisms (PSM). The two important groups of soil microorganisms that are used in commercial biofertilizer production include phosphorus solubilizing bacteria *Pseudomonas striata, Bacillus polymyxa, B. circulans, B. megaterium* var. *phosphaticum, Burkholderia cepacia,* etc. and phosphorus solubilizing fungi (PSF) *Aspergillus awamori, Penicillium digitatum,* etc. Their inoculation to different crops is found to be highly beneficial.

Procedure for mass production

(i) Desired PSM obtained from an authentic source is grown on Pikovskaya's medium in small (250 ml) flasks.

(ii) Inoculum from small flask is transferred to large flasks (1-2 litre capacity) and placed on a mechanical shaker for 3-5 days.

(iii) Under optimum conditions the growth of PSM attain a population of 10^8 or 10^9 cells/ml in 3-5 days

(iv) The culture obtained in flasks is called as starter culture and can be used for large scale production.

(v) Large quantity of liquid medium (broth) is prepared and is placed in large flasks or bottles or fermenter. Inoculum from the starter culture at the rate of 1% by volume is added and the culture is allowed to grow for 3-5 days.

(vi) The culture is then mixed with a carrier material and packed in polythene bags.

(vii) Lignite, charcoal and vermiculite are the common carrier materials.

Crops on which they are used

All cereals, pulses, oil seeds, plantation crops, vegetables, flower crops, fodder crops, commercial crops like sugarcane, tobacco, chilly, cotton etc.

Dosage, method and time of application

Seed treatment: seed treatment is done at the rate 400 g/ha. Method of seed treatment is same as that given under *Rhizobium* fertilizer.

Soil application: 25 kg/ha mixed in 20 kg FYM and can be applied by broad casting or by row method.

Slurry treatment/Root dip method: A thick slurry of 1 kg inoculum in 5 litre water is made and roots of seedlings are dipped in slurry for a minimum period of 30 minutes and then planted in the main field.

Inoculation to plantation crops: For plantation crops, the soil around the root zone is excavated in the form of ring. The PSM biofertilizer is applied all along the root zone in the form of a basin or ring @ 25-50 g per plant after mixing with well decomposed FYM / vermicompost. The excavated soil is then covered back around the trunk of the tree. Similarly, plantation crops, vines such as pepper can be inoculated.

Shelf life

Biofertilizer packets normally have a shelf life of six months at room temperature. During the end of this period each gram of the inoculant shall contain 10^7 cells per gram of the carrier (on dry weight basis).

Compatibility : Similar to *Rhizobium* fertilizer

Quality control specifications prescribed for phosphate solubilizing bacteria

(i)	Base	Carrier based* in form of moist/dry powder or granules, or liquid based
(ii)	Viable cell count	cfu minimum 5 x 10^7 cells/g of powder, granules or carrier material or 1 x 10^8 cells/ml of liquid.
(iii)	Contamination level	No contamination at 10^5 dilution.
(iv)	pH	6.5-7.5 for moist / dry powder, granulated or carrier based and 5.0-7.5 for liquid based
(v)	Particle size in case of carrier based material	All material shall pass through 0.15-0.212 mm IS sieve.
(vi)	Moisture percent by weight, maximum in case of carrier based	30-40%
(vii)	Efficiency character	The strain should have phosphate solubilizing capacity in the range of minimum 30%, when tested spectrophotometrically. In terms of zone formation, minimum 5 mm solubilization zone in prescribed media having atleast 3 mm thickness.

* Type of carrier

The carrier material such as peat, lignite, peat-soil, humus, wood charcoal or similar material favouring growth of the organism.

Marking on packets should be Similar to *Rhizobium* biofertilizer

Do's and Don'ts for entrepreneurs, dealers and farmers

Similar to *Rhizobium* biofertilizer.

Cyanobacteria (Blue Green Algae)

Cyanobacteria are free-living photosynthetic nitrogen fixing microorganisms. They can use sunlight as the source of energy for their growth. They grow in aquatic and marshy environments, hence are best suited as biofertilizer for rice crop. Cyanobacteria vary in shape and the size of each cell is about 2-3 μm. The colour is often green or blue-green, but may appear purple, brown or red under certain environmental conditions. The site of nitrogen fixation in algal filament is the heterocyst, which does not have photosynthetic pigments. *Nostoc, Anabaena, Tolypothrix* are some of the common nitrogen fixing cyanobacteria. In addition to fixing atmospheric N, they also produce plant growth promoting substances and improve physico-chemical properties and organic content of the soil. Unlike bacteria which need exogenous energy to fix atmospheric nitrogen, these cyanobacteria convert solar energy to energy required for nitrogen fixation. These cyanobacteria can thus be used as biological input in rice fields.

The cyanobacteria that are generally used as biofertilizer in rice are species of *Aulosira, Tolypothrix, Scytonema, Nostoc, Anabaena, Plectonema,* etc., as composite culture. Nitrogen fixed by cyanobacteria becomes available to the associated plant either by exudation or their mineralization after the death of these organisms. In paddy fields, the death of cyanobacterial biomass is associated with drying cycles and their mineralization is rapid due to drying and wetting cycles. The agronomic trials conducted with rice crop commonly show that nitrogen contributed by cyanobacteria is of the order of 20-30 kg per hectare.

Procedure for mass production (small scale)

(i) Prepare shallow troughs (3′x 6′x 9″) of galvanized iron or permanent concrete tanks. More number of tanks can be prepared to produce more material.

(ii) Introduce 8-10 kg of sieved soil and mix well with 200 g of super phosphate. Add water up to 5-15 cm depending on the local conditions and rate of evaporation. The reaction of the soil should be around neutral. If acidic, correct by adding lime.

(iii) To prevent insects, add carbofuron (3% granules) at the rate of 20 g per trough or with other suitable insecticides.

(iv) After the soil has settled sprinkle the cyanobacterial culture on the surface of the standing water.

(v) In hot summer months, the growth of cyanobacteria will be rapid and in about seven to ten days they form a thick mat. If the daily rate of evaporation is high, add water to maintain required level. When the cyanobacterial growth becomes sufficient, stop watering.

(vi) Allow the water to evaporate completely so that the dried cyanobacterial growth cracks into flakes.

(vii) Collect the dried flakes from the troughs and store in bags for use in the field.

Pit method

Instead of troughs or tanks, shallow pits are dug in the ground and layered with a thick polythene sheet to hold the water. Other procedures are the same as in the trough method. This method is easy and less expensive to operate by small farmers.

Field method

The field production of cyanobacteria is really a scaled up operation of the trough or pit method to produce the material on a large scale.

(i) Demarcate the area in the field for cyanobacterial production; the suggested area is 40 m^2. If cyanobacterial production is envisaged immediately after crop harvest, the stubble is to be ploughed and if the soil is loamy it should be well puddled to facilitate water logging.

(ii) Bund the area with strong 15 cm sized bunds.

(iii) Flood the area with water to a depth of about 2.5 cm and apply lime. In the trough and pit methods, flooding is done only in the beginning; in the field, flooding may be repeatedly needed to keep the water standing.

(iv) Apply super phosphate at 12 kg/40 m^2. To control the predators like daphnids, snails and mosquitoes, apply Carbofuron (3% granules) or Ekalux (5% granules) at 250 g/40 m^2.

(v) Apply the cyanobacterial culture at the rate of 5 kg/40 m^2.

(vi) In clayey soils, good growth of cyanobacteria takes place in about two weeks in clear sunny weather, while in loamy soils it takes about three to four weeks.

(vii) Once the cyanobacteria have grown forming floating mats, they are allowed to dry and the dried cyanobacterial flakes are collected and stored in plastic bags for further use.

(viii) One can continuously harvest cyanobacteria from the same area by reflooding the plot and applying superphosphate and pesticides.

(ix) Addition of cyanobacterial inoculum for subsequent production is not necessary.

(x) During summer months (April-June), the average yield of cyanobacteria per harvest ranges from 15-30 kg per 40 m^2.

Dosage, method and time of application

Broadcast 10-20kg cyanobacterial flakes per hectare, 3-4 days after transplanting rice seedlings. Keep a little standing water in the field for about 10 days after inoculation. When nitrogenous fertilizers are used, reduce their dose by 1/3 and use cyanobacterial inoculum to get the yield equivalent to the full dose of the fertilizer. Avoid applying fertilizer and cyanobacterial inoculum at the same time.

Crops on which they are used: Wetland rice

Shelf life

The sun dried cyanobacterial flakes packed in polythene bags can be kept for a year without any loss in their viability.

Compatibility

Recommended pest control measures and other crop management practices can be undertaken without any detrimental effect on the cyanobacterial activity. But avoid mixing the inoculum with fertilizer or pesticides at the time of application.

Do's and Dont's for entrepreneurs, dealers and farmers

Similar to *Rhizobium* biofertilizer.

Azolla

Azolla is a free floating fresh water fern harbouring nitrogen fixing blue-green algae '*Anabaena*' in its leaf. There are different species of

Azolla, which include *A. pinnata, A. nilotica, A. mexicana,* etc. It is also used as a green manure, as it can double its biomass in just two days under favourable environmental conditions. The presence of *Anabaena* as symbiont in the dorsal lobe of the *Azolla* leaf can result in fixation of 60 to 80 kg N/ha/season. Within 20-30 days *Azolla* can multiply and produce biomass ranging from 15-25 tonnes per hectare, thereby adding enormous amount of organic matter to the soil, which improves soil fertility. It also reduces the weed problem in rice fields, as it multiplies faster then weeds. Besides its use as biofertilizer, *Azolla* is also used as poultry and cattle feed, as it contains high protein and mineral content.

Procedure for mass production

(i) Prepare 2 m^2 plots, 3 weeks before transplanting rice. Eight nursery plots are required for raising *Azolla* for inoculation in 1 hectare rice field.

(ii) Fill the plots with water to a depth of 10 cm.

(iii) For controlling insect-pests, add Furadan @ 0.02% to the nursery plots. Dissolve 10 kg fresh cattle dung in 20 litres water and sprinkle in each plot.

(iv) Broadcast fresh *Azolla* culture at 100 g in each plot, obtained from a reliable source.

(v) Add 30 g of single superphosphate at 4 days interval.

(vi) Maintain water in the plots continuously and observe for the growth of *Azolla.*

(vii) After 15-20 days the inoculum multiplies 100 folds, resulting in a yield of 40-50 kg fresh *Azolla* in each plot.

Crops in which they are used

Wetland rice and kitchen garden as green manure.

Dosage, method and time of application

Azolla from the nursery is added to the field 2-3 weeks before transplanting rice. Another method is to introduce *Azolla* a week after transplanting rice. The recommended rate of *Azolla* inoculum in the field is 10 tonnes/hectare. Application of *Azolla* before planting rice is more effective than introducing it after transplanting the crop. A

13-54% increase in rice yield over the control has been observed in *Azolla* inoculated rice plants.

Shelf life

The fresh *Azolla* inoculum is heavy and hence difficult for transportation to long distances. Storage value of fresh *Azolla* is also low. Thus, *Azolla* inoculum has to be maintained throughout the year at several centers. This also ensures a steady supply of the fern during the year.

Compatibility

Use of *Azolla* in combination with Carbofuran blended urea and neem cake blended urea have shown increase in grain yield of rice. Use of Furadan during mass multiplication reduces attack by snails, slugs and breeding of mosquitoes and other insects. Reduced amount of N fertilizer and recommended level of P and K can be used along with *Azolla*. Avoid use of fertilizers at the time of *Azolla* inoculation.

Do's and Dont's

(i) Do not allow the water to dry up in the plots, while growing *Azolla* culture.

(ii) Expose the nursery plot to the sun.

(iii) Do not store the prepared material in direct contact with chemical fertilizers or other agricultural chemicals.

Arbuscular Mycorrhizal Fungi

The most common association occurring in majority of the plants important in agriculture, horticulture and tropical forest trees is the arbuscular mycorrhizal type. Arbuscular mycorrhizal fungi are geographically ubiquitous and occur over a broad ecological range from aquatic to desert environments. Majority of the plants are colonized by arbuscular mycorrhizal (AM) fungi excepting those belonging to few families like Brassicaceae, Chenopodiaceae, Polygonaceae etc. Arbuscular mycorrhiza are formed by Glomeromycetous fungi, the common genera being *Glomus, Gigaspora, Acaulospora, Entrophospora,* and *Scutellospora*. These fungi are obligate symbionts and have not been cultured on nutrient media in the laboratory. Hence, they are maintained in the roots of living host plants as "pot cultures". It is very well established that inoculation with efficient AM fungi improves the growth and yield of crop plants. This is particularly so in soils deficient in nutrients, especially phosphorus. This is mainly because of increased surface area of absorption of nutrients by the hyphae. Other beneficial effects are better water uptake, control of root pathogens, production of plant growth regulators etc.

Arbuscular mycorrhizal fungi are not host specific. However, it has been shown to have host preference *i.e.* for every host there is always a best fungal partner. Thus after screening several AM fungi, efficient fungus for inoculating a particular crop has been developed. Because of the difficulty in mass producing AM fungi, the best way to utilize AM fungi for crop production would be to concentrate on crops which are normally grown on nursery beds or root trainers or polybags, where they could easily be inoculated with desired AM fungi, and then transplanted to the field. Nearly 25 to 50% of phosphatic fertilizer can be saved through inoculation with efficient AM fungi. Tissue cultured plantlets also respond very well to AM fungal inoculation. Co-inoculation of AM fungi with other beneficial soil organisms like N fixers, P solubilizers is more useful in improving the growth and yield of crop plants.

Procedure for mass production

Several methods like, spores of AM fungi as inoculum, alginate entrapped inoculum, nutrient film technique, expanded clay as carrier of AM fungi, aeroponic culture, root organ culture method etc. have all been tried to mass produce AM fungi, but the best method, at present, is to produce these fungi in association with the host roots. This substrate based method is described here.

Substrate based inoculum is commonly used for mass production of AM fungi. The main objective of this production technology is to produce maximum number of infective propagules in a short time, with least contaminants. The inoculum production should be done in a glass house. Large plastic troughs or cement cisterns can be filled with perlite : soilrite mix (1:1) by volume and inoculated with the starter culture of the desired AM fungus and sown with Rhodes grass as the host. If Rhodes grass is not available, Guinea grass or any other suitable grass can be used. Calcium ammonium nitrate (to give 80 ppm N) and rock phosphate (to give 10 ppm P) can be used as the N and P source. Ruakura nutrient solution can be added once in 8 days. Addition of the fungicide Captan or Rilon, the acaricide Furadan, and the insecticide, Sumicidin at half the recommended level can be mixed with the substrate to reduce the contaminants in the inoculum, with no effect on AM fungi. Plants can be harvested 75 days after sowing. Roots can be cut into small pieces, mixed with the substrate, air dried, packed in polybags and used as the inoculum. The inoculum must be free from root pathogenic organisms.

Crops on which AM inoculum can be used

AM inoculum can be used on any crop important in agriculture, horticulture and tropical forestry that is raised in nursery bed or root trainer or poly bag. Crops which do not respond to AM fungi inoculation are those belonging to Brassicaceae like mustard, cauliflower, cabbage, radish etc.

Dosage, method and time of application

Nursery application: About ¾ to 1 kg of inoculum is sufficient for treating the nursery bed of 1 m^2. The inoculum should be applied 2-3 cm below the soil. Sowing the seeds or planting the cuttings can be done after applying the inoculum.

Polybag nurseries: About 10 g of the inoculum can be applied around 6 cm below the soil surface and seeds can be sown 1-2 cm below the surface of the soil.

Inoculation for out plating: About 10 g of the inoculum is applied to the planting hole before planting the seedling or hardened tissue cultured plantlets.

Inoculation of established perennial crops: Already established fruit trees like papaya and crops like mulberry can respond well to mycorrhizal inoculation when AM inoculum is applied in the root region at the rate of 100 g of inoculum per tree/bush.

Shelf life

AM inoculum is made up of the substrate containing hyphal bits and thick walled chlamydospores, and root pieces containing mycelium, vesicles and arbuscules. The inoculum can be stored at room temperature for one year.

Compatibility

Most of the pesticicdes are detrimental to AM fungi. However, some fungicides like Captan and Rilon have no deleterious effect on

AM fungi. Similarly the acaricide cum nematicide Furadan, and the insecticide Sumicidin have no adverse effect.

Quality control specifications prescribed for AM Fungi

No quality control specification for AM fungi was available till recently. The quality control specifications prescribed very recently (Dec., 2010) is given below.

(i)	From/Base	:	Fine powder/tablets/granules/ root biomass mixed with growing substrate
(ii)	Particle size for carrier based powder formulations	:	90% should pass through 250 micron IS sieve (60 BSS)
(iii)	Moisture content per cent maximum	:	8-12
(iv)	pH	:	6.0-7.5
(v)	Total viable propagules/ g of product, minimu	:	100/g of finished product
(vi)	Infectivity potential	:	80 infection points in test roots/g of mycorrhizal inoculum used

Do's and Dont's for entrepreneurs, dealers and farmers

Similar to *Rhizobium* biofertilizer.

Attempts to culture AM fungi on artificial media have met with little or no success

Plant Growth Promoting Rhizomicroorgnisms

The bacteria and fungi isolated from the root zone capable of promoting plant growth are referred to as plant growth promoting rhizomicroorgnisms (PGPRM). PGPRM increase plant growth through production of phytohormones, improved nutrient uptake and/or suppression of plant disease. Though nitrogen fixing and phosphate mobilizing organisms are also PGPRM, since they have been discussed earlier, other organisms are discussed here. They are organisms like *Pseudomonas fluorescens* and *Trichoderma harzianum,* which are commercially available.

Procedure for mass production of *Pseudomonas fluorescens*

(i) Desired *Pseudomonas fluorescens* obtained from an authentic source is grown on nutrient broth in small (250 ml) flasks.

(ii) Inoculum from small flasks is transferred to larger flasks (1-2 litre capacity) and placed on a mechanical shaker for 4-5 days.

(iii) Under optimum conditions the growth of *P. fluorescens* attain a population of 10^8 or 10^9 cells/ml in 4 to 5 days.

(iv) The culture obtained in flasks is called as starter culture and can be used for large scale production.

(v) Large quantity of liquid medium (broth) is prepared and is placed in large flasks or bottles or fermenter. Inoculum from the starter culture at the rate of 1% by volume is added and the culture is allowed to grow for 4 to 5 days.

(vi) The culture is then mixed with a carrier material and packed in polythene bags.

(vii) Lignite, wood charcoal and vermiculite are the common carrier materials.

Procedure for mass production of *Trichoderma harzianum*

(i) Desired *Trichoderma harzianum* obtained from an authentic source is grown on potato dextrose broth in small (250 ml) flasks.

(ii) Inoculum from small flasks is transferred to larger flasks (1-2 litre capacity) and placed on a mechanical shaker for 4-5 days.

(iii) When there is a good growth of the fungus a thick mycelial mat is seen. Broth is shaken well to break the fungal mat and the culture thus obtained is called as starter culture and is used for large scale production.

(iv) Large quantity of liquid medium (broth) is prepared and is placed in large flasks or bottles or fermenter. Inoculum from the starter culture at the rate of 1% by volume is added and the culture is allowed to grow for 4-5 days.

(v) Then the broth along with the fungal mat is macerated to get a uniform suspension of the fungal mycelium and this suspension may have a population as high as 10^8 colony forming units.

(vi) The culture is then mixed with a carrier material and packed in polythene bags.

(vii) Lignite, wood charcoal and vermiculite are the common carrier materials.

Crops on which they are used

It can be used on all cereal crops like wheat, oats, maize, barley, sugar beet, cabbage, potato, cotton, vegetable crops and plantation crops.

Dosage, method and time of application: Similar to *Rhizobium* biofertilizers.

Shelf life: At normal room temperature they can be stored for 6 months.

Compatibility: Similar to *Rhizobium* biofertilizers

Quality control specifications prescribed for *Pseudomonas fluorescens* and *Trichoderma harzianum*

(i) CFU on selective medium minimum of 2×10^6 cfu per ml or g for *Trichoderma* and for *Pseudomonas fluorescens* it is 5×10^6/g or ml.

(ii) Pathogenic contaminants such as *Salmonella*, *Shigella* or *Vibrio* should not be present. Other microbial contaminants not to exceed 1×10^4 count ml/g

(iii) Moisture content should be between 8 - 12%.

(iv) Stability or cfu counts at 30°C and 65% RH.

Do's and Dont's for entrepreneurs, dealers and farmers

Similar to *Rhizobium* biofertilizers.

Mass Production Technology of Microorganisms used for Controlling Soil-borne Plant Pathogens

The PGPRs *Pseudomonas fluorescens* and *Trichoderma harzianum* are also effective biocontrol agents against soil-borne plant pathogens. The procedure used for their mass production has been given above. *Paecilomyces lilacinus* and *Pochonia chlamydosporia* are mass produced for the biocontrol of plant pathogenic nematodes. This is given below.

Paecilomyces lilacinus

Paecilomyces lilacinus is an antagonistic fungus strongly parasitic to eggs, egg-masses and females of plant parasitic nematodes. The mycelium of the fungus infects egg-masses and unhatched eggs and destroys egg contents and embryos in 5 days. It causes localized degradation of egg shell by secreting enzymes. Sometimes the fungal mycelia after parasitization of eggs in egg masses can also enter the adult females through genital pores. Fungal parasitization can destroy up to 90% of eggs and 70-80% of egg masses or cysts. It is particularly effective against the root-knot nematode (*Meloidogyne* spp); golden cyst nematode (*Globodera pallida* and *G. rostochiensis*); cyst nematode (*Heterodera* spp.); citrus nematode (*Tylenchulus semipenetrans*); burrowing nematode (*Radopholus similis*); and reniform nematode (*Rotylenchulus reniformis*).

Mass production

The fungus has to be sub cultured on potato dextrose agar medium and incubated under dark condition for one week at 28 ± 1°C. Later with the help of cork borer (7 mm) discs should be cut and inoculated into 500 ml conical flasks containing 250 ml of potato dextrose broth and incubated for 15 days at room temperature and this is used as the mother culture for inoculating PDB in the fermenter @ 2 ml / litre. After 7 days fermentation, the fungal mat can be harvested, macerated and mixed with talc powder (carrier) @ 50 g fungal suspension per kg of talc powder containing 1% carboxymethyl cellulose as the adhesive. *P. lilacinus* is commonly used as a wettable powder, which contains at least 1% w/w of the bioagent.

Quality control specifications prescribed for *Paecilomyces lilacinus*

(i)	Viable cell count	Minimum of 5 x 10^7 cfu/g.
(ii)	pH	6.5-7.5
(iii)	Moisture	8-12%
(iv)	Shelf life	4 months at room temp not exceeding 32°C.
(v)	Contamination	Pathogenic contaminants such as *Salmonella, Shigella* or *Vibrio* should not be present. Other microbial contaminants should not exceed 1 x 10^4 count /ml or g
(vi)	Entomopathogenic capability	By bioassay on plates or target insects.

Pochonia chlamydosporia (*Verticillium chlamydosporium*)

Pochonia chlamydosporia is a nematophagous fungus which is a wide spread and naturally occurring facultative parasite of root-knot nematode eggs and has shown potential as a biocontrol agent against these nematodes. This fungus was found to parasitize the eggs by hyphal penetration and prevent their hatching. Investigators have reported that it is effective in controlling root-knot nematodes on various crops. To achieve the desirable control, the colonization of root system by *P. chlamydosporia* was found to be important and this colonized bioagent infects the eggs produced by first generation and second generation nematodes. Methods have already been standardized for studying the growth and survival of this fungus.

Mass production

The method used for mass production is similar to that described for *Paecilomyces lilacinus* earlier.

Quality control specifications prescribed for *Pochonia chlamydosporia*

Similar to that given for *Paecilomyces lilacinus.*

Microorganisms in Conversion of Agricultural Wastes to Compost

Organic matter is one of the important resources to be applied to soil to maintain the microbiological processes and recycling of nutrients. The decomposition of the organic wastes provides nutrients for the microbial succession in general and for biofertilizers in particular. Presence of organic matter brings about various microbiological processes and biochemical activities facilitating the release of and recycling of nutrient elements that are very much required for plant growth. Crop residues and other agro-industrial wastes are the main resources for organic matter. These organic materials are obtained on-farm as well as off-farm. To bring about effective decomposition of the organic materials, it is best managed by adding efficient ligno-cellulolytic fungi and bacteria for biotransformation into simpler readily available nutrients. Fungi are more effective in degrading lignocellulose. The important fungi used are *Phanerochaete chrysosporium, Pleurotus* spp., *Trichoderma* spp., *Paecilomyces fusisporus* and *Trichurus spiralis.*

Procedure for mass production

Fungi are mass multiplied by two methods:

(i) Solid state fermentation

(ii) Liquid fermentation

Solid state fermentation

Mass production of fungus by solid-state fermentation is more efficient.

Preparation of the substrates: Growth medium is prepared by mixing sawdust, wheat bran and straw dust in the ratio of 2:1:1 (i.e., 50 kg of sawdust is mixed with 25 kg of wheat bran and 25 kg of straw dust). It is sufficiently moistened with water to get moisture content of about 50 to 60%. Filled in aluminium trays of 24″x 30″x 3″, covered by brown paper and sterilized by autoclaving. The solid substrates are ready for fungal inoculation.

Preparation of starter culture: Starter culture is prepared by inoculating the required fungus into 3 litre of potato dextrose broth in a 5 litre culture bottle and incubating on a shaker for 6 days to get good sporulation. The fungal culture suspension is inoculated at the rate of 3 litres per 100 kg of the substrate for mass production. Fungal suspension is poured on the sterilized substrate, covered with the brown paper and incubated for 7 days. The fungus growing on the surface and the substrate form the fungal inoculum which is packed in poly bags for distribution.

Liquid state fermentation

Preparation of the medium: Potato dextrose broth is a very convenient low cost medium for cultivation of fungi on commercial scale. The mass production can be under aerated conditions using fermenters or stationery condition using flasks. Three litres liquid medium is prepared in 5 litre flasks and sterilized by autoclaving. To this 100 ml of fungal spore suspension is added aseptically and incubated for 7 days to get good sporulated mycelial mat on the surface of the medium. The mycelial mat is harvested, macerated in a blender, and mixed with the broth on which the culture was grown. The fungal suspension is mixed with a suitable carrier material like lignite, talc or powdered compost at the rate of 1:3, moisture content adjusted to 60% and packed in poly bags (1 kg / bag) for distribution.

Precautions to be taken: Sterilized carrier material should be used. Moisture should be maintained at 60% for better survival. Microbial inoculants should be stored in cool place and in shade.

Dosage and method of application

The fungal inoculants are applied for faster lignocellulose degradation at the rate of 13 kg / ha in case of *in situ* decomposition.

For faster decomposition of organic wastes in composting heaps/pits/windrows the inoculum is added @ of 2-5 kg (in layers)/tonne of organic waste.

Shelf life: Six months from the date of manufacture

The suggested standards are:

Fungal growth during mass production must be maintained in pure culture. The colony forming units (cfu) at harvest should be 10^7 / g of carrier. After mixing with carrier material, the population should be 10^7/g of carrier. Name of the fungus should be indicated on the packet.

Mass Production Technology for Microbial Pesticides

There are several naturally occurring microorganisms or their products that hold promise for the control of major insect pests. Microorganisms that affect insects are termed entomopathogens. They may be used to induce diseases in target insects or to suppress populations of insects directly or in combination with chemical insecticides. By adopting a systems approach to "integrated pest management", using entomopathogens and other nonchemical factors for specific pests, developing countries have an important opportunity to bypass the traditional chemical approach to insect control. The pressing need for sustainable crop production has established the importance of microbial pesticides for the management of pests. Microbial pest control agents like *Bacillus thuringiensis, Beauveria bassiana, Metarhizium anisopliae, Nomurea rileyi, Verticillium lecanii, Paecilomyces* spp., nuclear polyhedrosis and granulosis viruses and a few species of entomopathogenic nematodes belonging to *Steinernema* and *Heterorhabditis* have been found useful in controlling economically important pests affecting agricultural and horticultural crops and forest tree species. While enormous amount of research has been conducted in testing the biocontrol potential of numerous isolates of these organisms in the laboratory and field, achievements at economic mass production on commercial scale and development of stable formulations with adequate shelf life, field stability and efficacy have been rather limited.

Entomopathogens are produced by fermentation methods or in living insects. Fermentation technology is used for some bacteria and fungi, whereas living insects can be used for the production of obligatory parasitic viruses and protozoa. Both processes have been successfully used to produce commercial entomopathogenic products. Entomopathogens are infectious, replicating living organisms that are a natural part of our environment. Evidence that microbial insecticides pose little human or environmental hazard has been demonstrated by laboratory animal testing data developed to support pesticide registration. Microbial insecticides, like chemical insecticides, are usually sprayed or dusted on crops. Mass production technology used for some of the important microbial insecticides are given below.

Entomopathogenic Fungi

Among the bio-agents used for insect control, considerable work has been done on entomopathogenic fungi. Production of fungi to be used as biocontrol agents aims at the maximum possible number of living propagules that can infect insects. This is the distinguishing feature between production of fungi as biopesticides and fungi for other products; the biomass produced as biopesticide needs storing, while in latter most of the processes use fresh, actively growing organism. It is expected that fungal propagules in biopesticides must be alive at the time of their use. *Beauveria bassiana, Metarhizium anisopliae,* and *Nomurae rileyi* are some of the entomopathogenic fungi that have been investigated extensively as biopesticides. Most of the entomopathogenic fungi can be grown on a variety of solid substrates. The solid substrates commonly used for mass production of entomopathogenic fungi are various cereal grains (sorghum, rice, wheat, maize, oats, barley) and agricultural by-products (rice bran, rice husk, puffed rice waste, rice flakes, wheat bran, maize meal, maize cobs, soy flour, soybean mash, groundnut hull meal, bengalgram husk, sugarcane bagasse, cane trash, potato, etc). These solid substrates are moistened, sterilized and inoculated with the fungus in bottles/ polypropylene bags and incubated at optimum temperature and humidity for a period of 10-20 days. Then the fully-grown substrates are either macerated or used directly for foliar spraying/soil application in the field.

Alternatively, talc based formulations can be prepared by homogenizing the substrate along with the fungal growth into a fine powder and mixing with inert materials like, talc (1:1 to 1:2 proportion) and used for foliar sprays. The spore concentration in the formulation should be more than $1x10^8$ spores / g or ml. The entomogenous fungus can also be mass produced in liquid media like molasses yeast medium, carrot yeast medium, coconut waste water, potato dextrose broth etc.

These fungi can be grown in static liquid cultures / shake cultures / fermenters. The fungal biomass along with medium are mixed with sterilized talc powder (1:1 ratio) and dried in trays in aseptic conditions.

Diphasic system involving liquid medium followed by growth in solid medium is also used for mass producing these fungi. In this method, the fungus is allowed to grow in liquid media (fermenters) up to the end of log phase for maximum production of blastospores cum mycelial biomass, then it is transferred to solid nutritious or inert substrates for production of aerial conidia. This is done because the blastospores are highly sensitive to high temperatures and other environmental stresses and their viability is lost quickly. The inoculum in the solid substrates can be formulated into spray fluid or dry preparations for field application. The most commonly used methods for the mass production of some of the fungal insecticides is given below:

Beauveria bassiana

Production on solid substrates

Beauveria bassiana could be cultured on crushed moist sorghum/ rice/maize grains. The pure culture of fungus is grown on potato dextrose agar (PDA) or Sabouraud's maltose agar slants in tubes for 10 days at 26°C and the fungus is suspended in sterile water + 0.15% Tween-80. The spore load in the suspension is adjusted to 10^6 / ml by adding sterile distilled water. The grains are crushed in a mixer for 30 seconds to make small pieces, which can pass through 12-mesh size sieve. Then, 200 g crushed grains are placed in 0.2 mm thick 250g capacity polypropylene bags and 190 ml water is added. The bags are heat sealed and autoclaved at 121° C for 45 minutes. After cooling, 5 ml of spore suspension is added by cutting open one corner of bag under aseptic conditions. The opened area is resealed. Cotton plugging of the bags is preferred for proper aeration. The bags are incubated at 26°C for 15-20 days. The fungal mass along with grain carrier is harvested and dried at 40°C for 24 hours and ground in a mixture to get a fine powder. The powder is suspended in required water along with 0.02% Tween - 80 for preparation of spray fluid containing spore loads of 10^6-10^8 /ml. The spore suspension can be applied as foliar sprays in the cool and humid conditions (evenings). Dust formulations of the fungus can be developed by mixing the fungal preparation with talc for direct broadcast on leaves or as soil drenches.

Production on liquid media

Potato dextrose broth (PDB) can be used as a liquid medium for mass multiplication for *B. bassiana*. For this PDB can be prepared using 200 g of potato, 20 g of dextrose in 1,000 ml of water. Five hundred ml of this medium is taken in 1000 ml flasks and autoclaved at 120°C for 20 minutes. Each of these flasks is inoculated with 5 ml of spore suspension (1 x 10^8 conidia/ml) of *B. bassiana*, grown on PDA or Czapek's agar and incubated at 25°C for 15 days. After 15 days, the fungal growth on PDB is collected along with the medium and blended in the mixer for 1 to 2 minutes to get a homogenous slurry. This contains 10^9 conidia / ml. Five hundred g of talc powder (100 mesh) is sterilized in trays in hot air oven for 3 hours. After cooling, 500 ml of the homogenized fungal slurry is mixed with 500 g of sterilized talc powder in the trays (1:1 ratio). Then, this mixture is dried in laminar flow hood for 2 to 3 days under aseptic conditions (8% moisture) and packed in sterilized polypropylene bags. This formulation prepared from PDB contains 10^8 to 10^9 conidia/g. Five to ten g of formulation is mixed with 1 litre of water along with 0.02% Tween-80 and sprayed in the evenings.

Note: Large scale production can be done using fermenters. Molasses yeast broth can be used instead of PDB.

Metarhizium anisopliae

Procedures given above for the mass production of *B. bassiana* can be followed for the mass production of *M. anisopliae* also.

Quality control specifications prescribed for *B. bassiana* and *M. anisopliae*

(i) The cfu count should be minimum 1 x 10^8/ml or g (during the entire shelf life).

(ii) Pathogenic contaminants such as *Salmonella, Shigella,* or *Vibrio* should not be present. Other microbial contaminants not to exceed 1 x 10^4 count/ml or g.

(iii) Moisture content should be between 8-12%.

(iv) Stability or cfu counts at 30°C and 65% RH.

Entomopathogenic Bacteria

Bacillus thuringiensis (*Bt*) is one of the best-known and most widely used microbial control agents. It is pathogenic for lepidopteran larvae affecting more than 150 larval species that include some of the most important economic pests. The entomopathogenicity of *Bt* has been largely or completely attributed to the parasporal crystal inclusions, also known as insecticidal crystal proteins (ICP's), cry proteins or as δ-endotoxins. *Bt* is conventionally produced on a commercial scale by liquid fermentation in stirred tank reactors.

(i) Desired *Bt* obtained from an authentic source is grown in nutrient broth in 500 ml flask under shake culture conditions for about 12 hours to get 10^8 cells/ml.

(ii) When the organism is in the mid-logarithmic phase inoculum from small flask is transferred to seed fermenter or large flasks. The medium used in seed fermenter is desalted soyflour 25 g, glucose 40 g, corn steep liquor 25 g and polypropylene glycol (antifoam) 1 ml per litre with pH 7.0 and maintained at 30°C with aeration.

(iii) When the bacterium is in its mid-logarithmic growth (usually around 8-12 hours) it is transferred from seed fermenter to production fermenter containing the same medium used in the seed fermenter, set at 30°C with aeration. The fermentation is complete when the broth contains 90% free spores and crystals (spore / crystal ratio < 1) and 10% vegetative cells.

(iv) The broth is cooled to 5-10°C, pH adjusted to 4.1 using 5 M sulphuric acid, centrifuged at >8000 g and ultrafiltered to give a slurry with a solid content of 12-15% and spray dried to a powder to have 6-8% moisture.

(v) The primary powder is formulated into a wettable powder for final use. The formulation method usually followed is primary powder (active ingredient) 20%, kaolin or clay (filler) 75.25%, silica powder (free flow agent) 0.75%, bevaloid (dispersant) 2% and surfynol (wetter) 2%.

(vi) The shelf life is 3 years when stored at 30°C or lower.

It can be used very effectively to control the diamond back moth *Plutella xylostella*, the American bollworm *Helicoverpa armigera*, the tobacco caterpillar *Spodoptera litura*, Bihar hairy caterpillar *Spilosoma oblique*, castor semilooper *Achaea janata*, etc. on a variety of crops such as cotton, red gram, soybean, tobacco, castor and vegetables and on the gypsy moth *Lymantria dispar* damaging forest trees.

To register the formulation information/data pertaining to identity and biological properties of the organism, analytical methods, effects on human health (toxicology), non-target effects (eco-toxicology), residues on treated products etc. have to be provided.

Quality control specifications prescribed for *Bacillus thuringiensis:*

(i) cfu on selective medium should be minimum 1 x 10^8 /ml or g.

(ii) Pathogenic contaminants such as *Salmonella, Shigella* or *Vibrio* should not be present. Other microbial contaminants not to exceed 1 x 10^4 count/ml or g.

(iii) Moisture content should be between 8-12%.

(iv) Stability of cfu bacterial counts at 30°C and 65% RH.

Baculoviruses

Virus diseases have been described for most major arthropod pests. About 700 viruses have been isolated from insects and mites. The NPV and GV, because of their specificity, safety, virulence, and stability, are probably the most promising candidates for development as viral insecticides. Insect viruses are strict parasites and must be mass-produced in living hosts or cell cultures. This means rearing the target pest insect and producing the virus by artificial infection, harvesting the virus, and formulating an effective insecticidal preparation. During early phases of the development of viral insecticides, insects were collected in the field and fed with contaminated foliage. Dying insects were than processed into virus preparations for subsequent use. Later developed techniques and semi-synthetic diets permit year-round virus production. Insect viruses should be tested to determine the effects they may have on man, other animals, and plants.Its use can then be monitored for several years to ensure that unforeseen environmental effects do not develop. In the light of present knowledge baculoviruses appear to be safe and specific.

Maximum research work has been conducted on the use of NPVs for tackling two major polyphagous pests – *Spodoptera litura* and *Helicoverpa armigera.* The leaf eating caterpillar *Spodoptera litura* which is polyphagous pest is also known as tobacco caterpillar. It is a serious pest of tobacco nurseries. *S. litura* is also a sporadic pest of cauliflower, cabbage, castor, cotton, groundnut, potato and lucerne. It has been recorded from over 100 cultivated and wild host plants. It causes serious losses to several crops when outbreaks occur. *S. litura* NPV is host specific and infects *S. litura.* It can be successfully multiplied on its host and utilized in the field. It is necessary to multiply *S. litura* larvae continuously for production of *Sl* NPV.

The culture of *S. litura* is initiated by collecting eggs from the fields of castor, cauliflower, lucerne, tobacco etc. The egg masses are sterilized by dipping in 10% formalin for 30 minutes, washed in running water for 30 minutes, dried on blotting paper and kept for hatching in sterilized glass vials (20 g capacity) or plastic boxes (7.5 x 25 cm). The freshly laid eggs can also be surface sterilized in 0.05% solution of sodium hypochlorite for 5 minutes. These eggs are washed several times in running tap water to remove the traces of sodium hypochlorite. After 3 days the newly hatched larvae are transferred to bouquets of castor leaves (the leaves are surface sterilized with sodium hypochlorite solution, rinsed in sterilized water and dried before making a bouquet) and kept in a plastic container containing sand at bottom. The pupae are collected 3 days after all the larvae enter the sand. Pupae are collected and kept in a plastic container for adult emergence and further multiplication. *S. litura* larvae can also be reared on a chickpea based semi-synthetic diet (composition given below).

Item	Quantity
A. Chickpea (Kabuligram) flour	105.00 g
A. Methyl para-hydroxy benzoate	2.00 g
A. Sorbic acid	1.00 g
A. Streptomycin sulphate	0.25 g
B. Agar-agar	12.75 g
C. Ascorbic acid	3.25 g
C. Yeast tablets	25 tablets
C. Multivitaplex	2 capsules
C.Vitamin E	2 capsules
10% Formaldehyde solution	2.00 ml
Distilled water	780.00 ml

390 ml of water is mixed with fraction 'A' of the diet in the blender which is run for two minutes. Fraction 'A' and 'C' are mixed and the blender is run again for 1 minute. Fraction 'B' is boiled in the remaining 390 ml water, added to the mixture of A and B and the blender is run for a minute. Formaldehyde solution is added in the end and the blender is again run for a minute.

Production of *Spodoptera litura* nuclear polyhedrosis virus (*Sl* NPV)

From the stock culture of *S. litura,* 90% of 7-9 day old larvae (4th instar; head capsule width 1.1 mm) are used for *Sl* NPV production and the remaining 10% for continuation of laboratory culture. The larvae are collected and starved for 8 hours. *Sl* NPV suspension [10^8 POBs (polyhedral occlusion bodies) / ml] is prepared in 250 ml of water (with teepol) in bottle. The larvae are exposed to *Sl* NPV infection by dipping the clean castor leaves in *Sl* NPV suspension, drying them under shade for 15-20 minutes and providing them to the larvae for two consecutive days. Thereafter, the larvae are fed on healthy (not treated with *Sl* NPV) leaves (the petiole of which is dipped in water for maintaining the leaf turgidity and freshness) for the remaining part of their life. Fresh leaves are provided every day for the larvae. The larvae could also be exposed to *Sl* NPV infection by transferring them on to semi-synthetic diet treated with *Sl* NPV suspension.

The virosed larvae show characteristic symptoms within 4-5 days of infection. They start dying from 7th day onwards and are placed at 300 per container containing drinking water and are allowed to putrify for 3 days. The *Sl* NPV infected larvae could be easily distinguished by the pinkish colour on the under surface of their skin which turns to white with the accumulation of POBs (on death of infected larvae). The skin ruptures and the white liquefied body contents ooze out. The larvae are ground and filtered through muslin cloth. The POBs are dried over calcium chloride or by acetone precipitation and formulated by adding permitted spreaders/wetting agents.

Application

NPV is mixed with water (100 ml of NPV in 200 litres water), 1% crude sugar and 0.01% teepol and sprayed in evening hours as the virus is photosensitive. Addition of UV protectants like tinopal/ranipal will be useful. First spraying should be as soon as the young larvae are observed and repeated 2 more times at fortnightly intervals. Since it is a stomach poison, spray should cover entire plant parts. Though baculoviruses are compatible with most of the chemical insecticides and fungicides except methylparathion, in general contact time should not exceed 12 hours. Boric acid at varying concentrations has shown not only to be compatible but also in enhancing in case of NPV of *Spodoptera frugiperda*. It is compatible with parasitoids, predators, and other biopesticides.

Granulosis viruses

Granulosis viruses are reported mainly from Lepidoptera. They are effective against *Agrostis segetum* attacking cotton, maize, tobacco and vegetables; *Chilo infuscutellus* in sugarcane, *Pieris rapae* in brassicas, *Cydia pomonella* in apple, *Erinnyis ello* in rubber, *Cnephasia* sp. in forest trees. The virus multiplied in the host, is crushed and purified by centrifugation to get a stock solution of 2.74 x 10^{10} capsular occlusion bodies/ml. The stock solution is diluted 10 times with water and mixed with talc to make a slurry along with a wetting and dispersing agent like Dedenol Bx and dried to form a powder. This wettable powder is used for spraying.

Quality control specifications prescribed for baculoviruses

Viral unit: NPV - 1 x 10^9 POB/ml or g

GV - 5 x 10^9 capsules/ml or g

Contaminations: *Salmonella, shigella* or *Vibrio* should be absent. Other microbial contaminants should not exceed 1 x 10^4 count/ml or g.

Note

1. In case of NPV, POB count will be taken with Haemocytometer and in case of GV shallow depth counting chamber will be used for counting the capsules.
2. Biological assay: Results of bioefficacy tests by diet surface contamination method against second instar host insect should be generated for determining the LC_{50} of the formulation as against standards.

Entomopathogenic Nematodes

Entomopathogenic nematodes in the families Steinernematidae and Heterorhabditidae have been known since 1929 and 1975, respectively, but they became commercially available only during 1992. The biology of *Steinernema* and *Heterorhabditis* has many similarities. Only the third stage infective juvenile (dauer larva) can survive outside an insect host and move from one insect to another. The infective juvenile carries the symbiotic bacterium in its intestine (*Xenorhabdus* for steinernematid and *Photorhabdus* for heterorhabditid). It releases the bacterial cells after entering the insect's haemocoel by way of natural openings (spiracles, mouth, anus) and, with *Heterorhabditis* directly through the soft cuticle of certain insects. The bacteria proliferate, kill the insect host (usually in 24-72 hours) and render its interior favourable for nematode development.

One or more generations develop until the host-derived nutrients are depleted. At this time, the nematodes synchronously become third-stage infective juveniles that leave the cadaver to forage for new hosts by detecting insect excretory products. At temperatures ranging from 18-28°C, the life cycle is completed in 6-18 days, depending on the host insect and nematode species. Breakthroughs in production and formulation have made commercialization of nematode products possible. Major accomplishments are:

(i) *In vitro* production of nematodes in numbers sufficient for field applications, at a cost generally competitive with chemical pesticides.

(ii) Consistent production of high-quality (*i.e.* viable, pathogenic) nematodes.

(iii) Development of nematode formulations that provide a shelf-life sufficient for storage and transport to the site of use.

(iv) Formulations that make application rapid and simple.

Steinernematid and heterorhabditid nematodes can be reared *in vivo* in insect hosts or they can be mass-produced *in vitro* on solid or liquid medium. EPN- steinernematids and heterorhabditids are obligate parasites of insects with third stage infective juveniles (IJs) serving as potential agents for control of insects with their mutually associated bacterium. *Galleria mellonella* is the most used host all over the world because it can be easily reared, very susceptible and an excellent host for nematode reproduction. (Hussaini *et al.*, 2008b).

Considerable work has been done on *in vitro* production on solid substrate and in liquid media. For solid culture, substrates such as beef or pork kidney and liver, chicken offal homogenate, wheat flour medium, egg yolk medium, modified dog biscuit medium, ground veal pulp, Wout's medium etc. have been tried and the best medium was modified dog biscuit medium. The method is labour-intensive and is particularly well suited for cottage industry where labour is plentiful. *In vivo* production is low cost technology, with low startup costs, require less capital and technical expertise and resulting nematodes quality is high, yet cost efficiency is low. The optimum temperature for mass production is 25-30°C resulting in 10^6-10^7 IJs/flask. Mass production of EPN by liquid fermentation method has also been standardized. The liquid fermentation process, is highly efficient for several steinernemid species but not for *Heterorhabditis* and is economical for industrialized countries.

The cost of mass production is a major constraint for nematode commercialization. Agitation of media is also a constraint as it would damage the nematodes. The shelf life of the product is also limited (2-6 months at 4-10°C). The immediate need is to develop a simple, efficient mass production system for insecticidal nematodes, subsequently simple carriers to formulate the product, mainly increased shelf life is needed. With these drawbacks EPN is mass produced and used against insect pests. Formulations with IJs partially desiccated, trapped in water-dispersible granules, consisting of silica, clays, starch, offer several advantages including shelf life up to 6 months at 20°C over other formulations.

Technological improvements in production, formulation, packaging and shelf life of the products may change the scenario. Since nematode products are safe to apply and do not contaminate the environment, some clients opt for a biological control method even at a higher cost. Also, at least in some situations, the nematodes become established; recycle and their offspring continue to control the target insect. Thus, the higher short-term cost may be lower in the long run when continued control by the recycling nematode is obtained. Thus, by the judicious use of nematodes and chemicals, it may be possible to reduce the cost of control and protect the environment at the same time.

References

Bagyaraj, D.J. 2004. Quality control and constraints in biofertilizer production technology. In: *Biofertilizers Technology for Rice Based Cropping Systems,* (eds. S. Kannaiyan, K. Kumar and K. Govindarajan). Scientific Publishers, Jodhpur, India. pp. 402-408.

Bateman, R. 2005. Case studies and context for mycopesticide implementation: The development of products based on *Metarhizium anisopliae*. In: *Microbial Biopesticide Formulations and Application*. (eds. R.J. Rabindra, S.S. Hussaini, and B. Ramanujam). PDBC Pub., Bangalore, India. pp. 130-138.

Bisoyi, R.N. 2000. Constraints of biofertilizer production and remedial measures. In: *Biofertilizers Technology for Rice Based Cropping Systems,* (eds. S. Kannaiyan, K. Kumar and K. Govindarajan). Scientific Publishers, Jodhpur, India. pp. 50-54.

Burges, H.D. (ed.) 1988. *Formulation of Microbial Biopesticides, Beneficial Microorganisms and Nematodes.* Chapman and Hall, London, UK.

Campbell, R. and McDonald, R.M. 1989. *Microbial Inoculants of Crop Plants.* Society of General Microbiology, IRL Press, Oxford, UK.

Deshpande, M.V. 2005. Formulations and applications of mycopathogens. In: *Microbial Biopesticide Formulations and Application.* (eds. R.J. Rabindra, S.S. Hussaini and B. Ramanujam). PDBC Pub., Bangalore, India. pp.150-158.

Goel, A.K., Sindhu, S.S. and Dadarwal, K.R. 2001. In: *Recent Advances in Biofertilizer Technology* (eds. A.K. Yadav, M.R. Motsara and R.S. Chaudhuri). Society for Promotion and Utilization of Resources and Technology Pub., New Delhi, India. pp. 207-256.

Gosal, S.K., Gupta, R.P., Pandher, M.S. and Kalia, A. 2001. Biofertilizers - potentials and constraints. In: *Applied Microbiology.* (ed. P.C. Trivedi). Agrobios India, Jodhpur, India. pp. 1-10.

Hazarika, L.K. and Puzari, K.C. 2005. Entomopathogenic fungi in rice pest management. In: *Microbial Biopesticide Formulations and Application.* (eds. R.J. Rabindra, S.S. Hussaini andB. Ramanujam). PDBC Pub., Bangalore, India. pp. 139-149.

Narayanan, K. 2005. Current status of baculovirus formulations and application in India. In: *Microbial Biopesticide Formulations and Application.* (eds. R.J. Rabindra, S.S. Hussaini andB. Ramanujam). PDBC Pub., Bangalore, India. pp. 27-40.

Pawar, A.D. 2005. Quality control and Registration requirements of microbial pesticides. In: *Microbial Biopesticide Formulations and Application.* (eds. R.J. Rabindra, S.S. Hussaini and B. Ramanujam). PDBC Pub., Bangalore, India. pp. 228-263.

Prasad, R.D. 2005. Formulations and application of *Trichoderma* species for biological control of plant diseases in India. In: *Microbial Biopesticide Formulations and Application.* (eds. R.J. Rabindra, S.S. Hussaini and B. Ramanujam). PDBC Pub., Bangalore, India. pp. 187-202.

Rabindra, R.J. 2005. Current status of production and use of microbial pesticides in India and the way forward. In: *Microbial Biopesticide Formulations and Application.* (eds. R.J. Rabindra, S.S. Hussaini and B. Ramanujam). PDBC Pub., Bangalore, India. pp. 1-12.

Rabindra, R.J., Hussaini, S.S. and Ramanujam, B. 2005. *Microbial Biopesticide Formulations and Application.* PDBC Pub., Bangalore, India.

Ramasamy, S. 2000. Automated mechanized systems in biofertilizer production: MFLs experience. In: *Proceedings of Symposium on Current Status of Biofertilizer Technology and its Promotion in Karnataka, Tamil Nadu, Kerala and Pondicherry,*(ed. R.N. Bisoyi), RBDC Pub., Bangalore, India. pp. 35-37.

Sathiah, N., Rabindra, R.J., Palaniswamy, S., Jayant, K.P., Subramanian, S., Senthil Kumar, C.M., Rajasekharan, B. and Kennedy, J.S. 2005. Microbial Control of *Helicoverpa armigera.* In: *Microbial Biopesticide Formulations and Application.* (eds. R.J. Rabindra, S.S. Hussaini and B. Ramanujam). PDBC Pub., Bangalore. pp. 49-88.

Subba Rao, N.S. 1982. *Biofertilizers in Agriculture.* Oxford & IBH Co., New Delhi, India.

Venkataraman, G.S. 1981. *Blue-Green Algae for Rice Production, FAO Soils Bulletin 46.* FAO Pub., Rome, Italy.

Media Composition

1. Ashby's mannitol medium

Mannitol	- 10.0 g
Potassium dihydrogen phosphate	- 0.2 g
Magnesium sulphate	- 0.2 g
Sodium chloride	- 0.2 g
Calcium sulphate	- 0.1 g
Agar	- 15.0 g
Distilled water	- 1000 ml

Agar and glasswares should be free of nitrogen.

2. Czapek's agar

Sucrose	- 30.0 g
Sodium nitrite	- 2.0 g
Dipotassium phosphate	- 1.0 g
Magnesium sulphate	- 0.5 g
Distilled water	- 1000 ml
pH	- 7.0

3. Dobereiner's medium

Malic acid	- 5.0 g
KOH	- 4.0 g
$MnSO_4.H_2O$	- 0.01 g
$MgSO_4.7H_2O$	- 0.1 g
NaCl	- 0.02 g
$K_2 HPO_4$	- 0.5 g
$NaMoO_4$	- 0.002 g
$CaCl_2$	- 0.01 g
$FeSO_4.7H_2O$	- 0.05 g
Agar	- 1.75 g
Bromothymol Blue (0.5% alcoholic soln.)	- 2.0 ml
Distilled water	- 1000 ml
pH	- 6.5-7.0

4. Jensen's agar

Sucrose	– 20.0 g
K_2HPO_4	– 1.0 g
$MgSO4.7H_2O$	– 0.5 g
NaCl	– 0.5 g
$FeSO_4$	– 0.1 g
Na_2MoO_4	– 0.005 g
$CaCO_3$	– 2.0 g
Agar	– 15.0 g

Distilled water – 1000 ml

5. Malt agar medium

Glucose	– 20.0 g
Malt extract	– 5.0 g
Peptone	– 1.0 g
Agar	– 20.0 g
Distilled water	– 1000 ml
pH	– 6.0

6. Modified Melin-Norkran's Medium

Ammonium phosphate dibasic	- 0.25 g
$CaCl_2$	- 1% solution (5ml aliquot)
$FeCl_2$	- 1% solution (1.2 ml aliquot)
Magnesium sulphate	- 0.15 g
Potassium sulphate	- 0.5 g
$NaCl_2$	- 1% solution (2.5ml aliquot)
Malt extract	- 2 g
Glucose	- 5 g
Agar	- 15 g
Distilled water	- 1000 ml

7. Modified dog biscuit medium

Dog biscuit	- 20 g
Peptone	- 0.5 g
Yeast extract	- 0.3 g
Beef extract	- 5.0 g
Oil	- 7 g

Mix all the above ingredients to make homogenous mixture. The polyurethane foam chips (1 cm^3) (2 g foam chips: 6-8 g medium)

8. Molasses yeast broth

Molasses	- 30 g
Yeast extract	- 5 g
Distilled water	- 1000 ml

9. Nutrient Agar

Peptone	- 5 g
Beef extract	- 3 g
Agar	- 15 g
pH	- 7
Distilled water	- 1000 ml

10. Okon's medium

K_2HPO_4	- 6.0 g
KH_2PO_4	- 4.0 g
$MgSO_4$	- 0.2 g
NaCl	- 0.1 g
$CaCl_2$	- 0.02 g
NH_4Cl	- 1.0 g
DL malic acid	- 5.0 g
NaOH	- 3.0 g
Yeast extract (Difco)	- 0.1 g
$FeCl_3$	- 10.0 mg
Na_2MoO_4	- 2.0 mg
$MnSO_4$	- 2.1 mg
H_3BO_3	- 2.0 mg
$Cu(NO_3)_2$	- 0.04 mg
$ZnSO_4$	- 0.24 mg
Bromophenol blue (0.5% in ethanol)	- 2 ml
Distilled water	- 1000 ml
Agar	- 15 g
pH	- 6.8

11. Potato dextrose agar

Peeled potato	- 250.0 g
Glucose	- 20.0 g
Agar	- 15.0 g
Water	- 1000 ml
pH	- 6.0 - 6.5

12. Pikovskaya's medium

Glucose	- 10 g
Tricalcium phosphate	- 5 g
Ammonium sulphate	- 0.5 g
Potassium chloride	- 0.2 g
Magnesium sulphate	- 0.1 g
Manganese sulphate	- trace
$FeSO_4$	- trace
Yeast extract	- 0.5 g
Agar	- 15 g
Distilled water	- 1000 ml

13. Sabouraud's dextrose / maltose agar

Polypeptone or neopeptone	- 10 g
Dextrose (or maltose)	- 40 g
Agar	- 15 g
Distilled water	- 1000 ml
pH	- 5.6

14. Waksman's No:77 medium

Dipotasium phosphate	- 0.5 g
Magnesium sulphate	- 0.2 g
Sodium chloride	- 0.2 g
Manganese chloride	- trace
Ferric chloride	- trace
Distilled water	- 1000 ml

15. Yeast extract mannitol broth

Yeast extract	- 1 g
Mannitol	- 10 g
K_2HPO_4	- 0.5 g
$MgSO_4.7H_2O$	- 0.2 g
NaCl	- 0.1 g
Distilled water	- 1000 ml
pH	- 6.5 - 7.0

Colour Plates

Chapter 3: Nitrogen Fixing Microorganisms

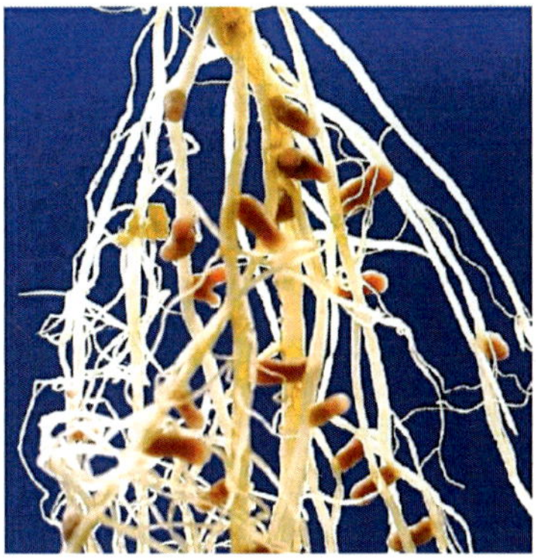
a

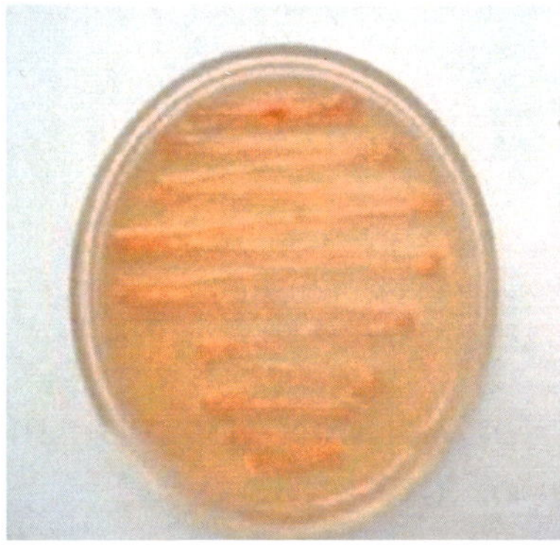
b

Plate 1 : a. Nodules of rhizobia in legume root, b. Rhizobia in plate

a

b

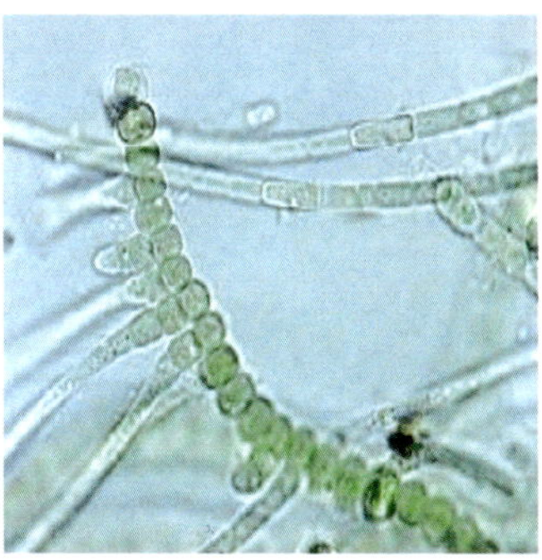
c

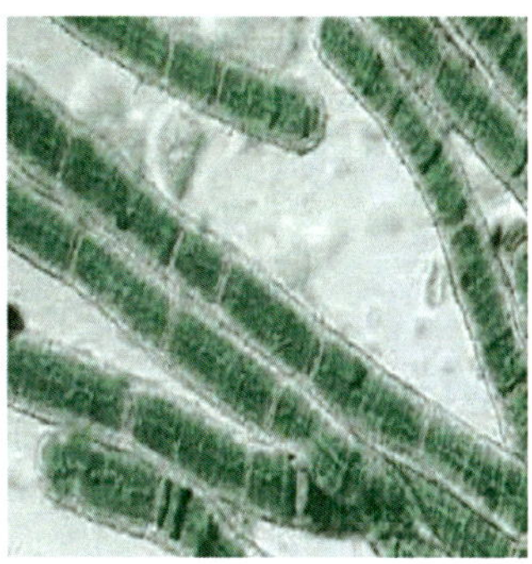
d

Plate 6 : a. *Nostoc* b. *Rivularia* c. *Fischerella* d. *Scytonema*

Plate 7 : Response of soybean to rhizobial inoculation
U – uninoculated, I – Inoculated

Plate 8 : *Azolla*

Chapter 4: Microorganisms Modulating Phosphorus Nutrition of Plants

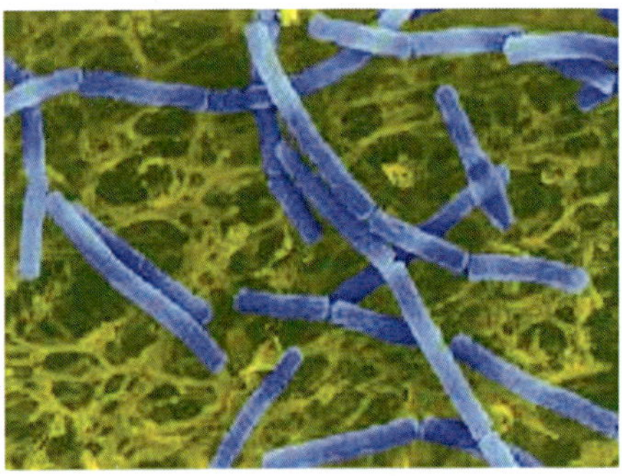

a

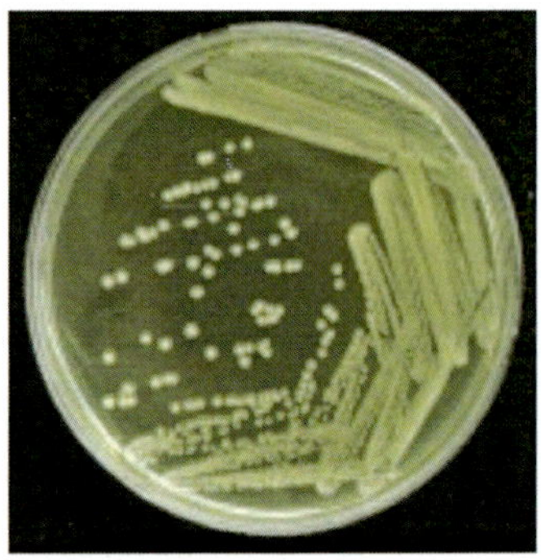

b

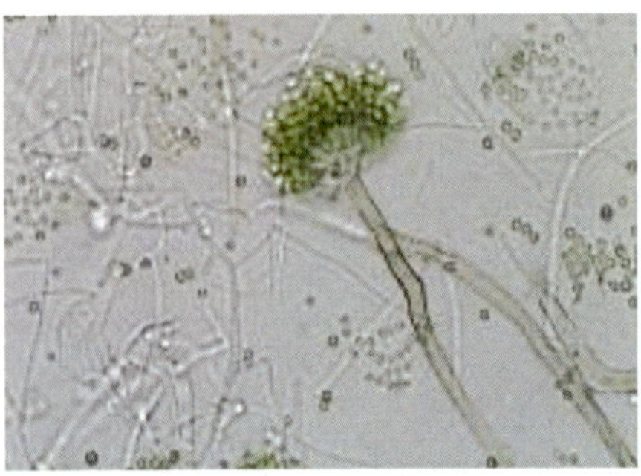

c

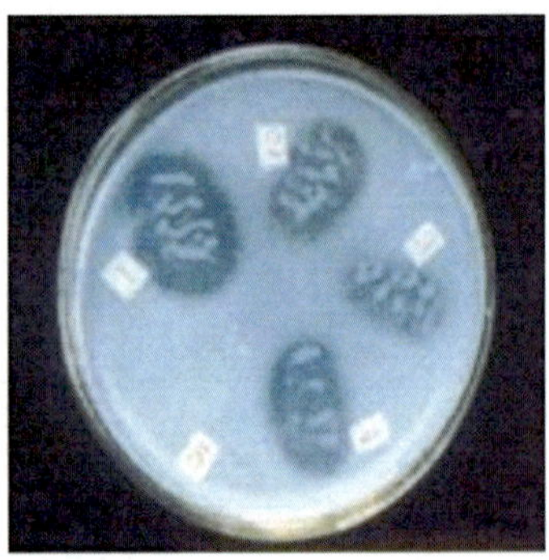

d

Plate 9 : a. *Bacillus megaterium* cells, b. Plate showing growth of *B. megaterium* c. *Aspergillus awamori*, d. Plate showing P solubilizing zones

Plate 10 : Ectomycorrhizal fungi. a. *Boletus pinophilus*, b. *Suillus cavipes*, c. *Russula decipiens*, d. *Laccaria laccata*, e. *Pisolithus tinctorius*, f. *Cenococcum* sp.

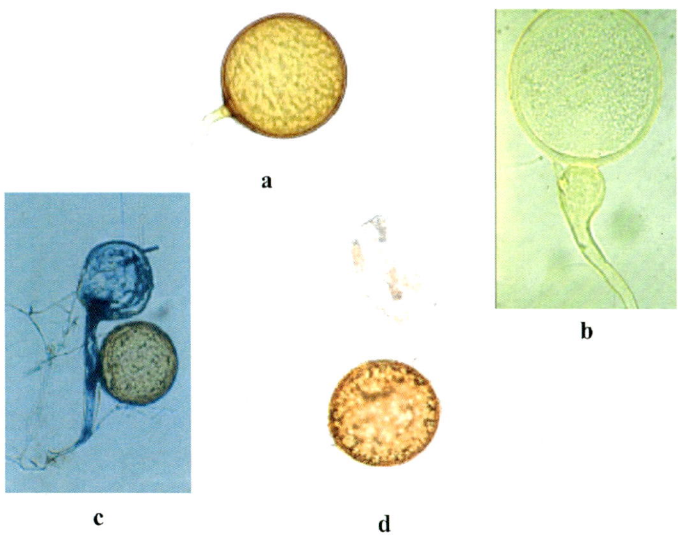

Plate 11 : Spores of AM fungi a. *Glomus,* b. *Gigaspora,* c. *Acaulospora,* d. *Entrophospora*

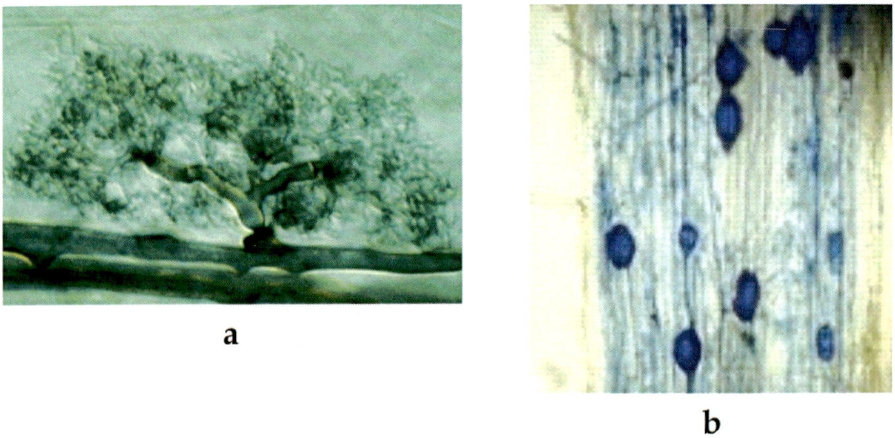

Plate 12 : AM fungal structures in root a. Arbuscule, b. Vesicles

Plate 13 : Response of plants to AM fungal inoculation a. Patchouli, b. Cardamom, c. Teak

a

b

Plate 14 : Response of a. *Piper longum*, b. *Acacia auriculiformis* to inoculation
U – Uninoculated, M – AM fungus, MC – Microbial consortia

a

b

Plate 15 : Growth of a. *Tectona grandis* b. *Acacia auriculiformis* planted in degraded forest 52 months after planting
U – Uninoculated; I – Inoculated with microbial consortia

Chapter 5: Plant Growth Promoting Rhizomicroorganisms

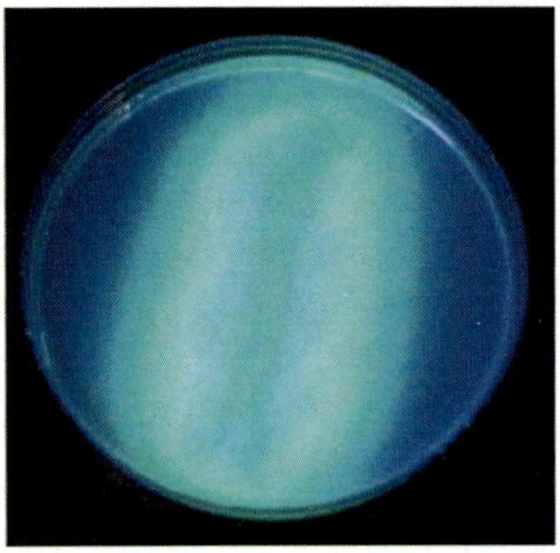

a

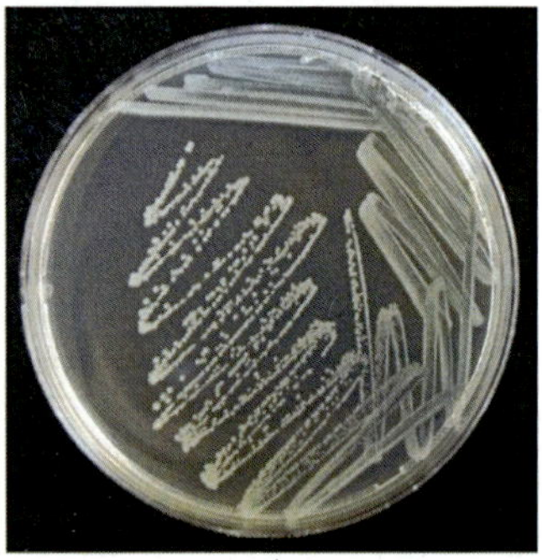

b

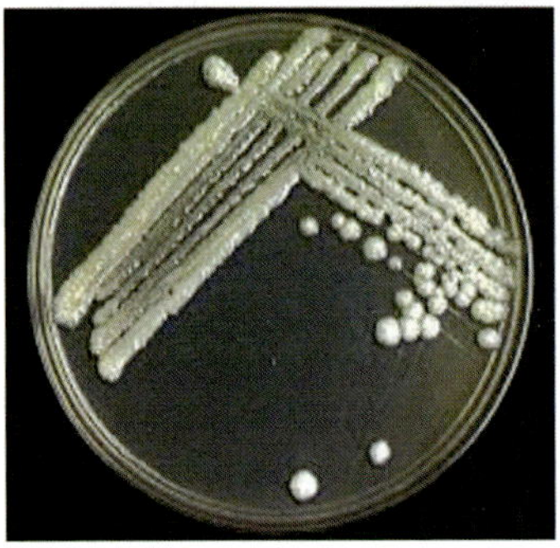

c

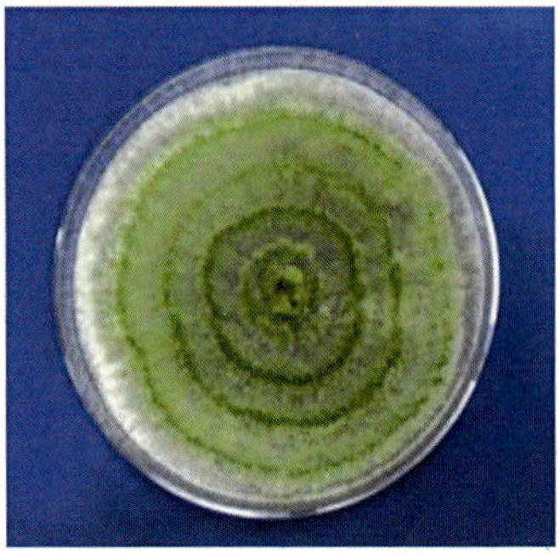

d

Plate 16 : Plates showing growth of PGPRs a. *Pseudomonas fluorescens,* b. *Pantoea agglomerans* c. *Paenibacillus polymyxa* d. *Trichoderma harzianum*

Chapter 7: Microorganisms for Biocontrol of Insect Pests

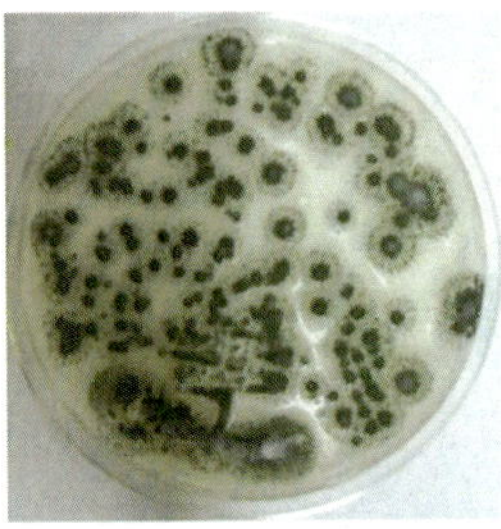

a

b

c

Plate 19 : Insect pathogenic fungus *Metarhizium anisopliae* a. Plate showing growth, b,c. Diseased insects

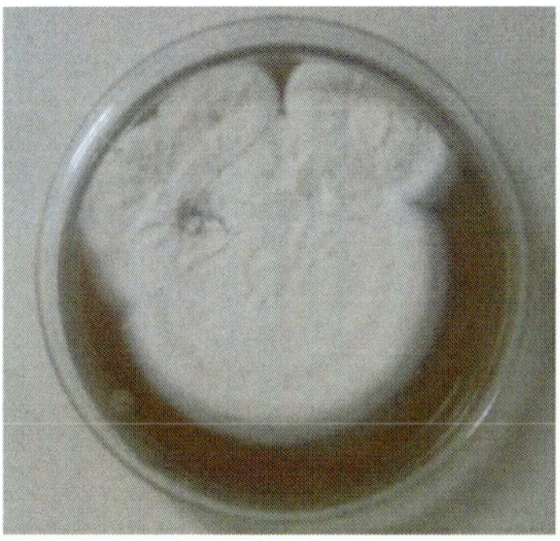

a

b

Plate 20 : Insect pathogenic fungus *Beaveria bassiana* a. Plate showing growth, b. Diseased insect

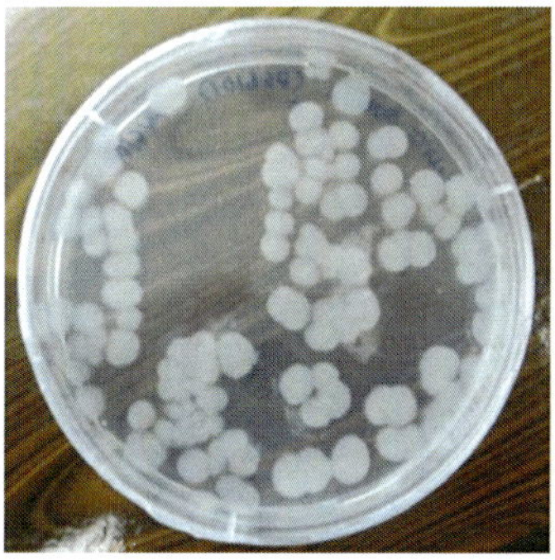

a

b

Plate 21 : Insect pathogenic bacterium *Bacillus thuringiensis* a. Plate showing growth, b. Diseased insect

Plate 22 : Insect diseased by NPV

Chapter 8: Microorganisms for Control of Weeds

Plate 23 : a. *Mikania micrantha* infected with *Puccinia spegazzinii* b. Spores of *P. spegazzinii*

Chapter 9: Microorganisms in Conversion of Agricultural Waste to Compost

a b c

Plate 25 : Fungi used for rapid composting a. *Phanerochaete chrysosporium*, b. *Trichurus spiralis*, c. *Pleurotus sajor-caju*